Handbook of Citizen Science in Conservation and Ecology

The publisher gratefully acknowledges the generous contribution to this book provided by the International Community Foundation/JiJi Foundation Fund.

# Handbook of Citizen Science in Conservation and Ecology

Edited by   CHRISTOPHER A. LEPCZYK, OWEN D. BOYLE,
and TIMOTHY L. V. VARGO

Foreword by   Reed F. Noss

UNIVERSITY OF CALIFORNIA PRESS

University of California Press
Oakland, California

© 2020 by the Regents of the University of California

Library of Congress Cataloging-in-Publication Data

Names: Lepczyk, Christopher A. (Christopher Andrew), 1970– editor. | Boyle,
    Owen D., 1974– editor. | Vargo, Timothy L. V., 1973– editor. | Noss,
    Reed F., writer of supplementary textual content.
Title: Handbook of citizen science in ecology and conservation / edited by
    Christopher A. Lepczyk, Owen D. Boyle, and Timothy L. V. Vargo ;
    foreword by Reed F. Noss.
Description: Oakland, California : University of California Press, [2020] |
    Includes bibliographical references and index.
Identifiers: LCCN 2019042776 (print) | LCCN 2019042777 (ebook) |
    ISBN 9780520284777 (cloth) | ISBN 9780520284791 (paperback) |
    ISBN 9780520960473 (ebook)
Subjects: LCSH: Conservation of natural resources—Citizen
    participation—Handbooks, manuals, etc. | Ecology—Handbooks, manuals, etc.
Classification: LCC S944.5.C57 H36 2020 (print) | LCC S944.5.C57 (ebook) |
    DDC 639.9—dc23
LC record available at https://lccn.loc.gov/2019042776
LC ebook record available at https://lccn.loc.gov/2019042777


28   27   26   25   24   23   22   21   20
10   9   8   7   6   5   4   3   2   1

To the citizen scientists of the past, present, and future

# CONTENTS

# CONTRIBUTORS

TEIS ADRIAN
NORDECO
DK-1159 Copenhagen, Denmark

EMMA ALBEE
Schoodic Institute at Acadia National Park
Winter Harbor, ME 04693, USA

HEIDI L. BALLARD
School of Education
University of California, Davis
Davis, CA 95616, USA

LORIANNE BARNETT
National Coordinating Office
USA National Phenology Network
Tucson, AZ 85721, USA;
School of Natural Resources and the
    Environment
University of Arizona
Tucson, AZ 85721, USA

ANDREW BEAHRS
Reef Check Foundation
Long Marine Laboratory
Santa Cruz, CA 95060, USA

RICK BONNEY
Cornell Lab of Ornithology
Ithaca, NY 14850, USA

ANNE BOWSER
Woodrow Wilson International Center for
    Scholars
Washington, DC 20004, USA

OWEN D. BOYLE
Bureau of Natural Heritage Conservation
Wisconsin Department of Natural
    Resources
Madison, WI 53703, USA

NEIL D. BURGESS
UN Environment Programme World
    Conservation Monitoring Centre
Cambridge CB3 0DL, UK

JENNIFER CALLAGHAN
Urban Ecology Center
Milwaukee, WI 53211, USA

REBECCA CHRISTOFFEL
Christoffel Conservation
Madison, WI 53704, USA

PEDRO DE ARAUJO LIMA CONSTANTINO
REDEFAUNA—Rede de Pesquisa em
     Biodiversidade, Conservação e Uso da
     Fauna Silvestre
Brasilia, DF, Brazil

ALYCIA W. CRALL
National Ecological Observatory Network
Boulder, CO 80301, USA

THERESA M. CRIMMINS
National Coordinating Office
USA National Phenology Network
Tucson, AZ 85721, USA;
School of Natural Resources and the
     Environment
University of Arizona
Tucson, AZ 85721, USA

FINN DANIELSEN
NORDECO
DK-1159 Copenhagen, Denmark

ELLEN G. DENNY
National Coordinating Office
USA National Phenology Network
Tucson, AZ 85721, USA;
School of Natural Resources and the
     Environment
University of Arizona
Tucson, AZ 85721, USA

SHANNON DOSEMAGEN
Public Lab
New Orleans, LA 70117, USA

SCOTT EUSTIS
Healthy Gulf
New Orleans, LA 70112, USA

JAN FREIWALD
Reef Check Foundation;
Institute of Marine Sciences
University of California, Santa Cruz
Long Marine Laboratory
Santa Cruz, CA 95060, USA

RACHEL GOAD
Pennsylvania Natural Heritage Program
Western Pennsylvania Conservancy
Harrisburg, PA 17105, USA

STEVEN GRAY
Department of Community Sustainability
Michigan State University
East Lansing, MI 48824, USA

EMILY M. HARRIS
School of Education
University of California, Davis
Davis, CA 95616, USA;
BSCS Science Learning
Colorado Springs, CO 80918, USA

BETH FETTERLEY HELLER
Urban Ecology Center
Milwaukee, WI 53211, USA

PER MOESTRUP JENSEN
University of Copenhagen
Thorvaldsensvej 40
DK-1871 Frederiksberg C, Denmark

REBECCA JORDAN
Department of Community Sustainability
Michigan State University
East Lansing, MI 48824, USA

NICOLE KAPLAN
USDA-ARS
Rangeland Resources and Systems
     Research Unit
Fort Collins, CO 80526, USA

CHRISTOPHER A. LEPCZYK
School of Forestry and Wildlife Sciences
Auburn University
Auburn, AL 36849, USA

EVA J. LEWANDOWSKI
Wisconsin Department of Natural
   Resources
Madison, WI 53707, USA

STACY LYNN
Natural Resource Ecology Laboratory
Colorado State University
Fort Collins, CO 80523, USA

SUSANNE MASI
Chicago Botanic Garden, Plants of Concern
   (retired)
Illinois Native Plant Society
Algonquin, IL 60102, USA

DAVID MELLOR
Center for Open Science
Charlottesville, VA 22903, USA

NATHAN J. MEYER
University of Minnesota Extension
Cloquet, MN 55720, USA

ABRAHAM J. MILLER-RUSHING
National Park Service
Acadia National Park
Bar Harbor, ME 04609, USA

JESUS MUÑOZ
NORDECO
DK-1159 Copenhagen, Denmark

GREGORY NEWMAN
Natural Resource Ecology Laboratory
Colorado State University
Fort Collins, CO 80523, USA

SARAH NEWMAN
Natural Resource Ecology Laboratory
Colorado State University
Fort Collins, CO 80523, USA

PAMELA LARSON NIPPOLT
University of Minnesota Extension
Saint Paul, MN 55108, USA

REED F. NOSS
Florida Institute for Conservation Science
Siesta Key, FL 34242, USA

KAREN OBERHAUSER
University of Wisconsin-Madison
   Arboretum
Madison, WI 53711, USA

JESSICA L. ORLANDO
Urban Ecology Center
Milwaukee, WI 53211, USA

MICHAEL J.O. POCOCK
Centre for Ecology & Hydrology
Crowmarsh Gifford, Wallingford
Oxfordshire, OX10 8BB, UK

RICHARD B. PRIMACK
Biology Department
Boston University
Boston, MA 02215, USA

MICHELLE D. PRYSBY
Department of Forest Resources and
   Environmental Conservation
Virginia Tech
Charlottesville, VA 22902, USA

ANNE REIS-BOYLE
Bureau of Wildlife Management
Department of Natural Resources
Madison, WI 53703, USA

LUCY D. ROBINSON
Angela Marmont Centre for UK
    Biodiversity
Natural History Museum
London, SW7 5BD, UK

ALYSSA H. ROSEMARTIN
National Coordinating Office
USA National Phenology Network
Tucson, AZ 85721, USA;
School of Natural Resources and the
    Environment
University of Arizona
Tucson, AZ 85721, USA

HELEN E. ROY
Centre for Ecology & Hydrology
Crowmarsh Gifford, Wallingford
Oxfordshire, OX10 8BB, UK

RUSSELL SCARPINO
Natural Resource Ecology Laboratory
Colorado State University
Fort Collins, CO 80523, USA

SARA N. SCHAFFER
National Coordinating Office
USA National Phenology Network
Tucson, AZ 85721, USA;
School of Natural Resources and the
    Environment
University of Arizona
Tucson, AZ 85721, USA

JENNIFER L. SHIRK
Cornell Lab of Ornithology
Citizen Science Association
Brooklyn, NY 11217, USA

AMANDA SORENSEN
University of Nebraska-Lincoln
Nebraska CoOp Fish & Wildlife Research
    Unit
Lincoln, NE 68583, USA

ANDREA LOREK STRAUSS
University of Minnesota Extension
Rochester, MN 55901, USA

JOHN C. TWEDDLE
Angela Marmont Centre for UK
    Biodiversity
Natural History Museum
London, SW7 5BD, UK

ELIZABETH TYSON
Woodrow Wilson International Center for
    Scholars
Washington, DC 20004, USA

TIMOTHY L. V. VARGO
Urban Ecology Center
Milwaukee, WI 53211, USA

PATI VITT
Lake County Forest Preserve District
Libertyville, IL 60048, USA

JAKE F. WELTZIN
U.S. Geological Survey
Ecosystems West Branch Office
Fort Collins, CO 80526, USA

ANDREA WIGGINS
University of Nebraska Omaha
Omaha, NE 68182, USA

# FOREWORD

REED F. NOSS

The need for reliable scientific information and expertise to guide policy, management, and decisions of all kinds related to the environment has arguably never been greater. We live in a world where anti-science, anti-intellectual, and even anti-truth attitudes prevail at the highest levels of government. Scientists are ridiculed as elitist nerds out of touch with the "real world" of making and buying things to support endless economic growth. Indeed, they are seen as impeding progress through their dire warnings about such trivial things as global warming and species extinctions.

The people in power who promote these false and dangerous views were put there by citizens who have an insufficient or incorrect understanding of science. These attitudes and trends must be countered in a democracy by other citizens armed with factual knowledge and critical thinking skills. Perhaps professional scientists might most effectively play this role in society. Yet the number of trained scientists available to collect, analyze, interpret, and disseminate scientific data is orders of magnitude below the need. And professional scientists are not necessarily the most appropriate or effective environmental advocates. With a continued lack of significant state and federal funding dedicated to the ecological sciences, as well as a reduced interest in science among some private foundations and non-governmental environmental groups, the prospects seem bleak for generating and communicating the information needed to make intelligent decisions on critical environmental issues.

This is where citizen science steps in. Given the shortage of professionally trained scientists and of funding to support the salaries of such professionals in the field and laboratory, the need for trained participants (usually volunteers) to assist in the process of scientific research is enormous. Moreover, the very process of assisting in scientific research might help create a citizenry who understand the process of science and its value to society, as well as benefiting the scientists overseeing the research. In the context of ecology and conservation biology, the focus of this book, citizen science has great potential to enhance ecological literacy and promote a conservation ethic among many members of the public. As the numbers of these trained and ecologically enlightened citizens grow, we might reach a point where democracy works again for the greater good.

I must confess to being initially skeptical of citizen science. How could members of the general public not formally trained in the sciences be counted on to collect data in a rigorous fashion? How could they learn field identification and other necessary skills in a short amount of time? Could they be trusted to enter data accurately and maybe even analyze those data? Finally, is it fair to consider amateurs working on a scientific project "real scientists"? Perhaps it was unfair of me to have such reservations and ask such questions, since I myself participated in what are now called citizen science projects, such as Christmas Bird Counts and Breeding Bird Surveys, long before I was a professional scientist with advanced degrees. Not only was this work fun and educational, but I felt that it was contributing valuable information—and it was. Data from the volunteer-based ornithological projects in which I participated have been published in many peer-reviewed journals and have been extremely informative about the status and trends of bird populations. I have also seen how participants in these projects develop a keen interest in the conservation of bird species in the regions where they live and work—as well as a deeper appreciation of nature in general and of the importance of conserving it. These are surely among the most important outcomes of citizen science.

This multi-authored book, edited by Christopher Lepczyk, Owen Boyle, and Timothy Vargo, assuages my reservations about citizen science. Previous studies have shown that, when done right and with rigorous training, citizen science works. The data collected in citizen science projects are virtually as reliable as those gathered solely by professional scientists, and at greatly reduced cost. As the editors hasten to point out in the Introduction, however, citizen science projects must not be seen as simply a way to conduct research on the cheap. Significant resources are still required to conduct high-quality citizen science projects, and highly trained professional scientists are usually needed to design the studies, train and supervise the volunteers, and analyze and interpret the data. Otherwise, citizen science is indeed not science.

The present volume germinated at a symposium on citizen science organized by the editors at the 2008 meeting of the Ecological Society of America, then gradually grew from there. The various chapter authors all make important contributions. Case studies abound. The ability of citizen science to vastly expand the geographic scope, temporal scale, and sample sizes of scientific projects—as well as their interdisciplinarity—is convincingly demonstrated here. The chapters go into great detail on such topics as the history of amateur participation in science, what kinds of projects are particularly suitable for citizen science, project design and funding, recruiting and training of participants, pitfalls to avoid, health and safety of participants, ethical and legal issues, data collection and analysis, quality control, data management, reporting of results, how to evaluate success, retention of participants, and promoting and publicizing projects. Building a sense of community among volunteers and leaders is essential. The range of specific advice offered here, for anyone thinking about organizing and leading a citizen science project, is enormous. I can hardly imagine initiating a citizen science project without first reading this book!

Citizen science offers an opportunity to advance scientific research and help the public develop a better understanding of Earth and its inhabitants. It benefits individuals by actively educating them about the process of science and giving them a "hobby" that provides a strong sense of purpose. It builds communities of volunteers and leaders and fosters friendships among them. It builds awareness of environmental issues and creates advocates for conservation and for science in the service of conservation. We need more of it.

## PREFACE

The seeds of this handbook go back many years, to when the three of us first began working together on a citizen science project in Milwaukee, Wisconsin, with our good friend William Mueller. At that time, *citizen science* was not a widely used term and there was only a loose affiliation of practitioners engaged in it. As we began our research on the stopover ecology of migrating birds in Milwaukee's parks, we worked to recruit participants, train them, and develop a successful project. But we had relatively little guidance, and as we discussed our project, it became apparent that there was a need for communication about how to engage in citizen science. This need was the catalyst for a symposium on citizen science that we organized in 2008 for the Ecological Society of America Annual Meeting, held that year in Milwaukee.

During the time leading up to the meeting, we started discussing the idea of creating a handbook that practitioners, students, research scientists, and the layperson could use to develop and guide an ecological or conservation-focused citizen science project, based on our own experiences and those of our colleagues. However, finding the time to develop such a book proved difficult as we focused on our day jobs and as many life changes intervened. But over the years, we continued to have many great discussions about the need for a handbook as we watched citizen science start to really flourish. Finally, after many starts and stops, we plotted out a book and a plan to produce it.

Hence, the handbook you are holding is the culmination of many years of friendship, collaboration, and thoughtful discussions. As we moved about the country, changed jobs, and started and completed many different projects, we found time to have late-night phone calls to discuss the book, write various components, and edit the chapters. We could not have completed this task without the loving support of our families (Jean, Olivia, Isabel, Jessie, Henry, Cecilia, Anne, and Linnea), the thoughtful comments of the reviewers, and the commitment of the authors to produce their chapters. Likewise, we are indebted to Blake Edgar, Merrik Bush-Pirkle, Bradley Depew, Kate Marshall, Stacy Eisenstark, Robin Manly, and the rest of the staff at the University of California Press who supported our work and provided critical feedback.

We hope this handbook will serve you well in your own projects.

*Chris, Tim, and Owen*
*January 2020*

# Introduction

CHRISTOPHER A. LEPCZYK, OWEN D. BOYLE,
and TIMOTHY L. V. VARGO

*Citizen science.* These words convey a variety of ideas and meanings, depending on one's background and training. In fact, not too long ago these words often stirred controversy in the scientific community. Today, however, citizen science is a fully fledged field that has contributed a great deal to our understanding of the world and provided an opportunity for countless numbers of individuals to participate with, and become, scientists.

While citizen science is widely practiced today, it is important to keep in mind that it is not a replacement for professional science. Science will continue to need highly trained professionals who have a formal education and experience in the scientific process, which often requires years of apprenticeship (i.e., graduate school). Just as we need doctors and dentists to have training and professional experience prior to practicing medicine on their own, we need similar requirements for scientists. Training and experience are particularly important in applied science contexts, such as conservation biology and natural resource management, where ignorance can result in long-term or permanent environmental problems.

Another important point to keep in mind is that citizen science is not a license to use unpaid volunteers to collect data. When considering a scientific project, it is important to avoid implementing a citizen science approach simply to reduce costs or because a grant did not provide the desired level of funding. Likewise, it is important for nonprofit organizations, government agencies, and academics not to turn to citizen science simply because of a lean budget year or as a means to reduce costs. In other words, citizen science should not be viewed as free labor.

Well-designed citizen science not only aids in scientific discovery, but can serve to educate the participants. In fact, one large reason for considering a citizen science approach is to improve scientific literacy (or, as in the case of this book, ecological literacy). Many people who engage in citizen science gain not only an improved understanding of the system or project they work with, but an increased general interest in science. Thus, citizen science is something to consider for people of all ages as a way to increase their appreciation for, and understanding of, science. Indeed, as we will see in this handbook, citizen science is being used as an educational tool and as a means to engage in outreach. Cooperative Extension programs, for example, have used it to connect people to the world they live in.

Because citizen science offers a powerful approach to answering questions and educating the public, it is important that it be carried out correctly. Hence, the focus of this handbook is on the practice of citizen science as it relates to conservation and ecology, which includes conservation biology, applied ecology, basic ecology, and natural resource management. The book is written primarily for practitioners working for government agencies, non-governmental organizations, and academic institutions who are interested in beginning partnerships with the community or enhancing and improving existing citizen science projects. However, the book should hold value for anyone interested in citizen science and science education.

## WHAT THIS BOOK IS NOT

We have sought to cover a wide swath, but we acknowledge that no handbook is perfect or can address everything, particularly with regard to an actively innovating field like citizen science. Hence, there are a handful of topics that we do not cover in this handbook, but we will note them here, given their importance.

First and foremost is funding. As the old saying goes, there is no such thing as a free lunch, and this is true in citizen science. *All* projects require financial resources of some type (if only money to pay for gas or paper to record data). Even if at first you cannot identify the financial costs of a project, simply consider the time involved, the number of people, and the tools they will need and it should become clear that all projects have (and require) a financial component. How projects are funded varies greatly, from grants to online funding aggregators. However, the type and amount of funding needed often depend on the type of project. Furthermore, organizations that fund projects vary markedly in what they are looking to fund and what they require of the grantee. Ultimately, there is no single pathway to funding and most projects will need to have individuals dedicated to raising funds and applying for grants. In the end, if your project does not have the resources to be completed, then the project should not be pursued.

Second, this handbook does not explicitly focus on experimental design. Any scientific project needs to have a methodology, including a framework in which to test questions, if it is to be successful. However, given the vast array of questions and hypotheses that ecologists and conservation biologists address, and the many approaches available and statistical tools needed, a primer on experimental design would require more space than is feasible in a handbook like this. That said, we strongly recommend that at least one member of any team planning a citizen science project understands experimental design; this may mean seeking out a professional scientist who can assist with that aspect of the project.

Third, the focus of this book is general and not aimed at one subdiscipline or segment of ecology and conservation. However, while subdisciplines and fields are not specifically called out in the chapters that follow, the approaches described are relevant for all. A case in point is environmental justice. One key element of evaluating environmental justice questions is environmental monitoring, which is a common form of citizen science. Keep in mind that we are focused here on process and the value of citizen science, not on specific taxa, ecosystems, or subdisciplines.

Fourth, we initially had a placeholder for a chapter related to systemic imbalances of power in the world of science and the importance of focusing on justice and the dignity of each stakeholder. We soon realized that a single chapter on these extremely important issues would run the risk of trivializing them and that to address them properly would require a separate book. With this in mind, we encourage all practitioners to strongly consider the social effects of their work. In true collaborations, one side is not deemed more important and it should not be assumed that one side is going to control the budget or write the grants. Nor should one assume that citizen scientists are replacing professional scientists, but one should recognize that all parties bring important and necessary skills to the table.

Finally, this book is not a philosophical discussion on the pros and cons of citizen science or on what citizen scientists should be called or how they should be valued. We view many arguments about citizen science to be settled (e.g., data quality) and others as minor in relation to the larger value of the field. That said, we do tend to use the words *participant* or *citizen scientist*, rather than *volunteer*, to describe individuals who participate in projects, in order to avoid the view that citizen science is free.

## ORGANIZATION OF THE HANDBOOK

We have laid out the handbook in three main sections. The first provides an introduction to citizen science and a history of its emergence. This history is valuable in framing the larger view of the handbook and provides the context in which citizen science has grown and become useful. The second section is the heart of the book and focuses on the specifics of citizen science, from project conception to completion. The final section focuses on real-world examples of citizen science projects at scales from local to continental, in different ecosystems and locations, and on a variety of topics. The goal of this final section is to illustrate existing citizen science projects in conservation, ecology, and natural resource management that can serve as role models for other projects.

PART ONE

# BACKGROUND

# ONE

# What Is Citizen Science?

JENNIFER L. SHIRK and RICK BONNEY

The broad field, and history, of citizen science encompasses many different ways that members of the public can become involved in scientific work. In natural resource management contexts, people often think of citizen science as long-standing bird-count projects that now span a century and that have led to peer-reviewed publications about global trends. To others the term may evoke the contributions of amateur natural historians that established our understandings of many species, work that predates the term *science* itself (think Darwin, or even Aristotle). Still others may reference decades-long, volunteer water-monitoring traditions, or newer low-cost sensor networks, each of which enables communities to bring scientific data to the defense of environmental integrity in their neighborhoods and watersheds. A milestone was reached in 2014 when the *Oxford English Dictionary* included the term *citizen science* with a definition broad enough to embrace all of these traditions: *"Scientific work undertaken by members of the general public, often in collaboration with or under the direction of professional scientists and scientific institutions."*

Citizen science is real science—and therefore it reflects as wide a diversity of purpose and design as does the realm of more conventional science. Scientific work through citizen science may be intended for peer-reviewed publication or to inform management. Likewise, citizen science work may address timely problems or monitor long-term trends. A given citizen science research question may call for precision in data collection, or it may be better served by accuracy and statistical power. In every case, citizen science results are subject to the same system of peer review that applies to conventional science—in all cases, this review must be based on the intended purposes and merits of the research efforts.

Although acceptance of volunteer contributions to conventional science has been slow, ecologists and natural resource managers have been among the first to acknowledge the value of citizen science efforts. Thoughtfully designed citizen science projects produce reliable data that are usable by scientists, policymakers, and the public. Data from such projects have now been used in hundreds of scientific articles and have informed numerous natural-resources policy and planning efforts (Dickinson et al. 2010). Indeed, our current understanding of the natural environment is due, in large part, to the amateur scientist community's decades of dedication and expertise.

More recently, the advent of the Internet and mobile technologies has enhanced opportunities to collect, store, share, manage, and analyze vast amounts of data quickly and easily. As a result, scientific projects can now deploy large numbers of volunteers and can record huge volumes of observations in centralized databases that can be analyzed in near real time. Building on this opportunity, citizen science participation has become much more widely possible at both large and small scales, and thousands of projects now engage millions of participants around the world. In fields ranging from astronomy to zoology, citizen science is helping to answer many challenging questions—in ecological disciplines, in particular, it can help address issues that affect both the environment and our everyday lives.

## WHEN IS CITIZEN SCIENCE AN APPROPRIATE TOOL?

In order to maximize the potential for citizen science to achieve intended results, researchers and managers must decide whether work with volunteers is an appropriate approach to the research at hand. Citizen science projects can tackle major challenges in managing natural resources and the environment, such as species management, ecosystem services management, climate change adaptation, invasive species control, and pollution detection and regulation. Understanding the relative strengths of citizen science can help determine when it can provide advantages over conventional science. These strengths include the following:

> *Citizen science can often operate at greater geographic scales and over a longer time than conventional science.* Only volunteers can cost-effectively collect some types of data over sufficiently large areas and long enough periods of time to produce scientifically reliable and meaningful datasets. For example, eBird, a project of the Cornell Lab of Ornithology, collects five million observations of birds each month from locations across the globe. At maturity, a project's dataset can be geographically broad, locally dense, and temporally consistent, offering many powerful opportunities for analysis.

> *Citizen science can increase resolution of data collection.* Even at small geographic scales, citizen science can benefit from mobilizing a large number of observers in a given area. In addition to increasing observational power, citizen science can also enhance access to data in new places such as private lands, thereby filling in gaps in datasets. In some cases, projects have benefited greatly from volunteers collecting data when scientists are not typically present, such as during the Arctic autumn and winter.

*Citizen science can speed up and improve field detection.* Having many eyes on the lookout can help detect environmental changes (e.g., changes in the onset of spring through plant phenology), identify phenomena that require management responses (e.g., population declines, incidences of pollution, introduction of an invasive species), and monitor the effectiveness of management practices.

*Citizen science can improve data and image analysis.* Humans have the ability to recognize patterns and interpret large amounts of data as well as to detect individual diversity. Volunteers with no specialized training can outperform state-of-the-art algorithms in certain analytical tasks. Volunteers can also extract information from digitally collected primary data (e.g., images or audio) by identifying and recording secondary information (e.g., the abundance, behavior, and frequency or duration of various phenomena), tasks that are often difficult for computers. Finally, volunteers have additional knowledge and perspectives that can complement or improve professional scientists' interpretations of data and results, whether from local or traditional knowledge that can shed light on patterns or seeming anomalies, or by bringing attention to unappreciated social dimensions of management, livelihoods, or ecological processes.

*Citizen science can help refine research questions.* Participants in citizen science are affected by and observe local natural resources and the environment in their daily lives, so they can help improve the relevance of location-specific research questions and make them more useful to managers and local communities. A full understanding of natural resource and environmental issues often requires a holistic perspective, including human dimensions. Citizen science can help provide this holistic perspective and improve research relevance.

*Citizen science can help researchers appreciate connections between humans and their environment.* Citizen science is well suited for interdisciplinary collaboration, particularly for projects that include both natural and social dimensions. Natural resource and environmental managers increasingly address the social aspects of difficult ecological issues, such as managing wildfires in the wild-land-urban interface. By engaging local community members, citizen science can facilitate a shared understanding among managers, scientists, regulators, policymakers, volunteers, and others of the complex social dimensions of the natural systems where people live.

## WHEN AND HOW *NOT* TO DO CITIZEN SCIENCE

Some scientific inquiries are not appropriate for citizen science. It is often assumed that the most common factor limiting volunteer participation in a scientific project is the ability of trained volunteers to meaningfully contribute to the science. Some research questions, methods, and analyses do require specialized knowledge, training, equipment, and time commitments that can make citizen science inefficient or impractical as an approach. However, an increasing number of new citizen science projects—and associated data-quality studies—indicate that unexpected areas of scientific inquiry are successfully being advanced through citizen science.

As innovators continue to overcome more and more perceived barriers to volunteer-based research, it is difficult to outline any hard-and-fast rules about what will not work. Some general concerns are outlined by Michael Pocock and colleagues (Pocock et al. 2014). Additionally, any given research area will have case-specific constraints. We suggest that you proceed with caution if any of the following statements apply:

*Your primary goal is public education.* Citizen science can be an effective means of facilitating learning and can be utilized to great effect in both formal and informal educational settings. However, research suggests that people participate in citizen science because they want to make a contribution or a difference in the world. If your project is not grounded in purposeful research, expect participation to fall flat.

*There is significant concern about biasing decision making.* This is not a reason to avoid citizen science, but it is definitely a reason to be careful in project planning. Even professional scientists must proceed with caution—and ideally transparency—in circumstances where they are both conducting research and informing decision makers. Similar quality controls can be used for both citizen science and conventional science. These quality controls can include training, collection of duplicate samples, and post-data-collection analyses designed to identify outliers and biases in the data. Quality controls should be used in most, if not all, citizen science projects, even when volunteers are not involved in decision making.

*Participant interest is unknown.* Not all citizen science projects stimulate widespread or even sufficient public interest, from either curiosity or concern. Because interests vary, people are selective about participating in citizen science. For example, charismatic species such as wolves, bears, and certain birds receive—in general—more public attention (and support for public funding) than other species, including most plants. Similarly, study sites near tourist destinations and college campuses tend to receive more attention than those in urban and industrial areas. In addition, studies in small or remote communities may be of great local interest, yet the pool of potential participants for a citizen science project may be small. For certain taxa and ecological processes and for some biogeographic regions or geographic locations, citizen science could be difficult to sustain if large or ongoing datasets are required.

*Activities put volunteers, species, or habitats at risk.* Fieldwork can involve potentially hazardous conditions for volunteers. Such hazards may include physical dangers but can also include issues of privacy, such as public data revealing home addresses or patterns of activity. Citizen science also has the potential to draw unwarranted or unexpected attention to sensitive species, and to increase environmental impacts in remote areas. Some level of risk is unavoidable—be aware of the risks that volunteers, organizations, and landscapes will tolerate. Also be sure that volunteers are aware of any dangers to themselves and to species or habitats, and that policies and procedures are in place as safeguards. These may include liability policies, privacy policies, and mitigation techniques for sensitive data.

*Sampling frequency is high—or low.* The need for frequent sampling can limit the feasibility of citizen science. Few volunteers are able to devote extended periods of time to scientific projects. Extremely frequent (e.g., daily) sampling needs may therefore discourage participation and increase turnover. There can also be a mismatch between the availability of volunteers and the availability of managers or their staffs. For example, participants may be available primarily on weekends, when staff is unavailable. As a result, it may be difficult to recruit citizen science volunteers for certain projects. At the other extreme, infrequent (e.g., annual) sampling could make it harder to sustain collection of high-quality data, because participants may have to relearn even basic protocols. A successful sampling design for volunteers lies in between, where sampling frequency is just enough to keep participants well practiced and able to gather consistent data, but not so high as to become onerous and discourage participation.

*Skills needed do not align with skills available.* Running a citizen science project calls for an individual or team with a diverse skill set, ready to manage everything from protocol development to outreach to data management. Additionally, citizen science can increase research efficiency, but only if the benefits of having volunteer assistants are not outweighed by the task of training or recruiting skilled volunteers. Keep in mind, however, that most projects will require an investment in planning, vetting, and training early on that may take some time to show returns.

In short, consider the value of citizen science for answering the question at hand, in comparison to other approaches. Innovative research teams continue to adapt citizen science approaches to overcome potential pitfalls and succeed despite perceived odds, so where you or others see challenges, be careful not to overlook innovative solutions and possibilities. But do consider the value of adapting to challenges, and then proceed if the potential benefits seem worth the investment or at least worth a try. In general, if you have reasonable expectations that align with volunteer interests, and if training opportunities and research protocols are vetted and refined through pilot testing, you have a fair shot at a successful citizen science project.

## BEST-FIT OPPORTUNITIES FOR CITIZEN SCIENCE

Considering all the aforementioned points, citizen science may be most advantageous under the following circumstances.

### Volunteers Can Collect High-Quality Data

Sometimes volunteers need only minimal training to be able to collect high-quality data. For example, it may be easy to collect insects or make simple measurements, such as tree circumference, without extensive instruction or instrumentation. Volunteers can also collect data that require following elaborate protocols or developing certain specialized skills, such as in many water-quality monitoring programs. Research has shown that volunteers with proper training and guidance can accurately identify specimens at various taxonomic levels and accurately assess important population attributes, such as species abundance and distribution. Individual volunteers can even develop the skills to use sophisticated analytical instruments. However, projects that depend on numerous volunteers should not expect a sophisticated level of skill from all individuals, nor should volunteers be expected to undertake activities that require extensive training or certification. Generally speaking, the simpler the methods, the easier it is to engage volunteers in collecting high-quality data. Simple tasks also make it feasible to increase the number of contributors and easier to sustain collection of high-quality data. Organizations should also use data quality controls to identify questionable data and correct or discard them. The use of quality controls is relevant for all types of survey and assessment, whether implemented by volunteers or by professional scientists.

### Volunteers Can Address Unanswerable Questions

Participation by volunteers can make it possible to address questions that would be unanswerable in any other way. Public participation can be integral to the ability to collect, analyze, and interpret certain data. A major strength of citizen science is its ability to collect fine-grained information over broad areas and long periods and to process large amounts of data (such as images), simply because the number of volunteers may exceed the number of professionals (including researchers, faculty, and students) by several orders of magnitude. In some cases, volunteers can obtain data inaccessible to government employees, such as data on private lands or data related to individual activities such as hunting or fishing. When a rapid response is needed, such as to environmental disasters or sudden large-scale bird or fish die-offs, research efforts can benefit from the ability to swiftly mobilize large numbers of volunteers. A few types of studies that lend themselves well to citizen science include the following:

> *Monitoring studies* that assess patterns, in space and/or time, of one or more ecosystem components (e.g., Is this species here now? How many individuals of this species are here now?) or functions (e.g., Is this process happening now?). Data collection is standardized (the same for all sampling locations) and effort-controlled (data are recorded even if none are found—i.e., zeros count!).

*Process studies* that assess the impacts of factors (e.g., hazardous-fuel reduction treatments or pollution) on ecosystem components or functions (e.g., nutrient and water cycling). The researchers control the level and duration of the exposure, and there is a control (e.g., the status quo).

*Opportunistic and observational studies* that do not follow a strict design but are often deliberate in the subject and timing of observation. These studies can be useful because of the scale of the data collection, the rarity of the phenomena observed (e.g., a rare species or infrequent weather event), or the timeliness of the observations (e.g., collecting information for crisis response, such as after earthquakes or oil spills).

### Public Participation Meets Organizational Goals

Public participation in the scientific process can serve the organization's goals for public input and engagement and helps in decision making through the generation of both scientific knowledge and learning. Public input can help identify the most relevant questions that a scientific study is designed to answer and the best methods to carry out the study, particularly if the research is focused on an issue that affects or involves local people. If research is intended to affect natural resource management or environmental policymaking decisions, then public participation could aid in developing locally appropriate research questions and methods, particularly if the management or policymaking question requires understanding how human behavior interacts with ecological processes. In addition, local or traditional knowledge can reveal complementary ways of understanding local ecology, environmental trends, and threats to species and livelihoods. When the research problem is informed by the best available knowledge and by multiple perspectives, research questions and methods can be formulated that take into account the integrated social and scientific dimensions that managers and policymakers must address.

### When Citizen Science Works Best

In brief, citizen science works best when

- the project's aims are clearly defined and shared from the start;
- the research activities are beneficial to both scientists and participants;
- the project team includes expertise in all aspects, from data analysis to communication;
- evaluation is built into the project design, and there is a willingness to listen and adapt;
- small-scale pilot trials are undertaken as proof-of-concept and to improve procedures;
- the participants are meaningfully recruited and supported;

- the motivations and skill levels of all involved are understood;
- the participants feel they are part of the team;
- the project is an efficient and enjoyable way to gather and analyze the data; and
- the quality of the data is measurable.

Around these basic guidelines, researchers continue to open new possibilities and frontiers for citizen science. Citizen science can be done in almost any setting, but it takes a particular combination of skills, intentions, interests, and careful design to make it work (see box 1.1).

## WHAT *IS* CITIZEN SCIENCE?

*Citizen science, in all of its diverse forms, is a means of public engagement in scientific work that can expand the scope, reach, and impact of research.* Citizen science can provide opportunities to collect data at scales otherwise not feasible for professional scientists alone and can engage members of the public in compiling and using issue-relevant evidence to effect change. Taken together, these opportunities have the potential to bring new power to addressing major challenges in conservation and natural resource management, particularly those that require attention to both social and scientific aspects of a problem.

But citizen science is a complex undertaking, often demanding dedicated time and money as well as a willingness to understand what it means to engage with the public. In ecology in particular, with so much at stake, citizen science gives us the opportunity to reflect on and learn about the human dimensions of ecological research and environmental management, as well as about the science. Citizen science provides an opportunity to do more than hand off this task to educators or outreach specialists. If we, as researchers and managers, embrace the opportunity—and challenge—to listen and learn as well as interpret and share, we may be best prepared to fulfill the potential of citizen science to expand knowledge for science-based conservation.

## BOX 1.1 Not *Whether* but *How* to Do Citizen Science

In order to maximize the potential for citizen science to achieve intended results, it is important to understand how the structure of a project—particularly the ways volunteers are engaged—will affect the project's outcomes. Projects undertaken to address questions or issues that require large amounts of data to be collected over time and space are often *contributory* in their approach to participation. These contributory projects are typically top-down in design, with scientists controlling all aspects of research: determining the questions or issues to be addressed, designing data-collection protocols, processing and analyzing the data collected by public participants, and communicating the results. A contributory approach thus allows scientists to rigorously structure the procedures for data collection and submission. Meanwhile, a contributory approach can provide volunteers a chance to hone their skills and learn research techniques, but with limited opportunities to engage with scientists (in large-scale projects) or with the process of science, which limits deeper learning.

Alternatively, research opportunities can emerge in response to a need identified by community members. Such community-driven projects, called *co-created* or *community science,* typically focus on studying issues of pressing concern for local residents, such as environmental health or degradation. The co-created approach to citizen science is most relevant to projects addressing a specific environmental question or problem that will benefit from establishing a community- or volunteer-led monitoring scheme in which all parties have a stake in the outcomes. Such projects can involve significant volunteer commitment, and all parties must be willing to listen, plan, adapt, and reach consensus about mutually beneficial research goals and approaches. As a result, co-created projects can provide deep insights—for participants and researchers alike—into science and management processes, but they can also be difficult to operate at large geographic scales.

Despite these generalities, individual citizen science projects themselves cannot always be easily categorized. Some large-scale contributory projects have distinct goals for policy and management and can be leveraged in a co-created fashion by local communities. The global eBird dataset, for example, is accessed by the Nature Conservancy to determine areas of California's Central Valley that the Conservancy can pay farmers to flood during times of peak migration. Likewise, community science projects can yield outcomes with implications at broader geographic scales, such as by aggregating community-level data to look at statewide trends in water quality. Therefore, it is worth taking into account the significance of participant engagement when considering not just *whether* but *how* to employ citizen science as a tool to meet needs in ecological science, management, or policy (Bonney et al. 2009, Shirk et al. 2012).

# TWO

# The History of Citizen Science in Ecology and Conservation

ABRAHAM J. MILLER-RUSHING, RICHARD B. PRIMACK,
RICK BONNEY, and EMMA ALBEE

## WHAT IS CITIZEN SCIENCE? A HISTORICAL PERSPECTIVE

As described in the preceding chapter, *citizen science* refers to the engagement of nonprofessionals in scientific investigations—asking questions, collecting data, or interpreting results. In the past, a large portion of citizen science was done by expert amateurs who carried out scientific investigations independently of professional scientists, or in collaborations as peers—an approach now categorized as a collegial model of citizen science (Shirk et al. 2012). These citizen scientists, or expert amateur scientists, still exist and make important contributions to many fields, especially taxonomy and natural history (Costello et al. 2013).

Prior to the professionalization of science in the late nineteenth century, nearly all scientific research in North America and Europe was conducted by amateurs—that is, by people who were not paid as scientists (Vetter 2011a). Many of these individuals were pursuing research because of an innate interest in particular topics or questions (e.g., gentlemen scientists), as a part of religious or government activities (e.g., in their roles as bishops or record keepers), or as a part of vocations or subsistence (e.g., farmers, herbalists, hunters) (Porter 1978). Many amateurs were recognized experts in their field and conducted research indistinguishable from, and sometimes superior to, that done by most professional scientists of the time, although public perceptions of amateur scientists were complex and changed through time (Vetter 2011a).

As early as the seventeenth century and probably earlier, some of these amateur experts had recruited nonexperts to contribute natural history observations. For example, in the mid-eighteenth century, a Norwegian bishop created a network of clergymen and asked them to contribute observations and collections of natural objects throughout Norway to aid his research (Brenna 2011). This type of networking was a common way for early ecologists, such as John Ray and Carl Linnaeus, to collect specimens and observations from across the known world. Such contributions by nontrained scientists have helped to build some of the most valuable collections of animals, plants, rocks, fossils, artifacts, and other specimens worldwide.

Others who have collected information and data about the natural world in the past include farmers, hunters, and amateur naturalists. For instance, winegrowers in France have been recording grape harvest days for more than 640 years (Chuine et al. 2004), while court diarists in Kyoto, Japan, have been recording dates of the traditional cherry blossom festival for 1,200 years (Primack et al. 2009). In China, both citizens and officials have been tracking outbreaks of locusts for at least 3,500 years (Tian et al. 2011). In the United States, among the oldest continuous organized datasets are phenological records kept by farmers and agricultural organizations, documenting the timing of important agronomical events, such as sowing, harvests, and pest outbreaks (Hopkins 1918, Schwartz et al. 2012).

## THE CHANGING ROLE OF CITIZEN SCIENCE

During the professionalization of science over the past 150 years, the role of amateurs as peers or colleagues to professional scientists diminished. Amateur scientists still abound, as evidenced by the many naturalist clubs (e.g., bird-, insect-, mushroom-, and plant-focused groups) across the United States and by the growth in citizen science worldwide, but the major roles of amateurs in conducting research have changed somewhat. They began to fill two major roles within scientific pursuits (although they contribute in many other ways, too).

The first role involves projects that tackle ecological questions at scales that are simply not possible without a great deal of on-the-ground effort. Programs that follow a contributory or collaborative participatory model (see chapter 3), such as the North American Breeding Bird Survey, the U.S. National Weather Service's Cooperative Observer Program (NWS-COOP), the North American Bird Phenology Program, and lilac monitoring programs (the latter two are now part of the USA National Phenology Network; see chapter 14), have yielded national- or continental-scale datasets of biological and physical data that could not have been collected otherwise. In the late 1800s and early 1900s, the National Weather Service, for example, was tasked with gathering weather data that were critical to a variety of aspects of the U.S. economy, particularly agriculture, but was provided with only a limited budget to do so (Vetter 2011b). The agency therefore followed the example of weather bureaus in Europe by turning to volunteers who were broadly distributed throughout the country. The outcome of their work was one of the most important long-term datasets in North America, essential for agriculture, development planning, and assessment of recent climate change.

The second major role that citizen science has filled is in undertaking projects that professional scientists would not do on their own, whether because of the type of question or the place of study. For example, research scientists have incentives to study questions that advance knowledge of the field as a whole and to avoid projects that are too restricted in scope to be widely cited or of interest beyond a narrow audience. Thus, many local, place-based projects go uninvestigated by professional scientists and are carried out instead by local residents. A project may focus on finding causes of local problems, such as pollution, wildlife deaths, or pest outbreaks, and may also lead to management or policy solutions once the causes are found. For instance, the volunteer program Save Our Streams was founded in 1969 to monitor, protect, and restore streams in Maryland (Firehock and West 1995). The program has since been used as a model for a national program supported by the Izaak Walton League of America and has been widely recognized for its role in understanding and restoring streams throughout the United States. Volunteer programs aimed at tackling local issues have long existed across the country and continue to make important contributions to science and resource management today.

## ECOLOGICAL INSIGHTS FROM HISTORICAL CITIZEN SCIENCE

### Museums, Herbaria, and Other Collections

Historical collections of specimens, photographs, and similar records held at museums and other institutions around the world have yielded innumerable insights into ecology, evolution, and conservation biology. Professional scientists and untrained or self-trained amateurs contributed to the creation of these collections. Such collections have been used extensively to develop modern taxonomic systems of naming and classifying and to understand the dynamics of evolution and the distribution of species.

More recently, ecologists and conservation biologists have utilized historical collections to analyze shifts in the distribution and abundance of species due to land-use change, climate change, and other anthropogenic forces (Jetz et al. 2012, Pimm et al. 2014). For example, herbaria in the northeastern United States tend to have large numbers of specimens collected between 1870 and 1940. By using intensive field surveys to compare the abundance and distribution of herbarium specimens with the modern abundance and distribution of the same species, researchers can document the decline of rare and endangered species and the arrival and spread of non-native invasive species (Lavoie and Saint-Louis 2008, Lavoie et al. 2012). Similarly, by comparing the flowering dates of herbarium specimens with flowering records gathered by current observers, we can detect shifts in flowering times associated with a warming climate (Primack et al. 2004). Other biological phenomena, such as the adult phases of butterflies, dragonflies, bees, and moths, can be investigated using the same approach (Bartomeus et al. 2011, Brooks et al. 2014). Collections of dated photographs—in particular, old landscape photos—also represent an enormous resource for studying changes caused by climate change, land use, air pollution, invasive species, overfishing, and the impact of deer, cattle, and other large herbivores (Miller-Rushing et al. 2006, McClenachan 2009, Webb et al. 2010).

### Small-Scale Datasets: Individuals and Groups

Historical records held by individual amateur naturalists or groups provide ecologists potential insight for understanding long-term changes in ecosystems. Some of these naturalists are famous, like Meriwether Lewis and William Clark (of the nineteenth-century Lewis and Clark Expedition) or President Thomas Jefferson, but most are relatively unknown beyond their local communities. Generally, the people who made these historical observations were not intentionally participating in scientific projects, or were collecting data to address questions unrelated to their current use. However, their observations, particularly those for which the methods of collection and other metadata are well documented, provide key data for current scientific studies. Researchers have begun using these datasets to gain important insights into ecological responses to climate change.

One particularly noteworthy dataset is the record of first flowering dates, first leaf-out dates, and first arrival dates for migratory birds in Concord, Massachusetts, made by the famous writer and early environmentalist Henry David Thoreau between 1851 and 1858. Thoreau's observations are so valuable because they were collected from one well-defined place over many years, and he was able to accurately identify the large numbers of species involved. Later botanists (professional and amateur alike) continued his observations of first flowering times, and also recorded the abundance of different plant species in Concord. Likewise, a series of ornithologists and amateur birdwatchers recorded the first arrival dates of birds in later decades. These historical records, combined with modern observations, have been used to demonstrate that plant phenology is responding more strongly than bird phenology to warming temperatures (Ellwood et al. 2010, Primack and Miller-Rushing 2012). Furthermore, these data can reveal relationships between phenological changes and declines in the abundance of many native species, as well as increases in the abundance of different invasive species (Willis et al. 2008, 2010).

Many groups of amateur naturalists have also kept important ecological records. For instance, members of the Cayuga Bird Club in Ithaca, New York, have been recording the arrival dates of migratory birds each spring since 1903. Although the goal of the bird club is simply to preserve a record of sightings and arrival dates, their observations are a valuable source of data for ecologists studying the impacts of climate change on bird migration patterns and have shown that many species are arriving earlier than in the past (Butler 2003).

## Large-Scale Programs and Datasets

Many of ecology's most important and widely used datasets come from citizen science programs. For example, the aforementioned NWS-COOP has been collecting basic weather data across the United States since 1890. The results inform much of what we know about variability and directional changes in climate over the past 120 years (Karl et al. 2009).

Other datasets collected by citizen science programs provide the basis for many policy and management decisions. Perhaps the best example is the widespread involvement of volunteers in monitoring water quality across the United States. The water quality data they collected have frequently been used by management agencies to define baseline conditions, identify problems, and determine what management actions were needed. These water quality data are so important that many states have passed rules providing guidelines for their use in policy and management decisions, sometimes considering volunteer-collected data analogous to data collected by state agencies (Firehock and West 1995).

Many wildlife ecologists, agricultural scientists, and resource managers also rely on citizen science for critical data. For example, some of the best datasets describing migrations, population dynamics, phenology, and pest outbreaks were generated by citizen science programs such as the North American Breeding Bird Survey, the Christmas Bird Count, and extensive agricultural monitoring initiatives. These datasets have led to insights regarding bird population dynamics and conservation status (Bled et al. 2013) and factors that influence bird wintering ranges (La Sorte and Jetz 2012). Furthermore, these large-scale datasets have contributed to insights that are now central to our understanding of ecology and agriculture, such as the recognition of the relationships among temperature, latitude, elevation, and plant and insect phenology (Hopkins 1918). Beyond monitoring programs, wildlife management agencies also rely on hunters for data on wildlife diseases, such as chronic wasting disease in deer, to improve our understanding of disease prevalence and transmission and to guide management and policy responses (Williams et al. 2002). Likewise, long-term records kept by both amateur and commercial fishermen are increasingly being analyzed to detect changing patterns in the abundance and structure of fish populations and to determine whether fisheries management is having the desired outcome (Rosenberg et al. 2005, Granek et al. 2008, McClenachan 2009).

## DISCOVERING AND ANALYZING PAST CITIZEN SCIENCE DATA

Interest in climate change, land-use history, invasive species biology, and conservation has led many researchers to search for historical records gathered through citizen science methods. Such records include observations of species occurrences, population sizes, behaviors, and phenology, as well as community-level records, such as local floras and faunas, gathered by individuals or groups of people with specialized interests. Although such datasets can be challenging to find, an unexpectedly large number exist and are being organized and archived by government agencies and research institutions, such as the USA National Phenology Network, iDigBio, and the National Park Service, which hold records dating back to the 1700s or earlier.

Historical datasets of all types must be interpreted carefully because we do not always know how the data were gathered (Lepage and Francis 2002). Additionally, because many of these historical datasets have been collected by different observers over time, determining whether patterns or trends in the data are genuine or the result of changes in observer or methods can be challenging. For instance, a variable as simple as the number of days per week that two observers gathered data could substantially affect the results (Miller-Rushing et al. 2008). Nevertheless, we have found a surprising number of historical citizen science datasets that are documented reasonably or very well and include high-quality data (Primack and Miller-Rushing 2012).

## PROGRESS IN CITIZEN SCIENCE THROUGH HISTORY

### Innovations

Over time, the techniques involved in developing and managing citizen science projects have changed, improving both the scientific and educational outcomes of many projects. Advances in communications, transportation, and computing have made it easier for volunteers to contribute and for scientists and volunteers to manage and analyze the resulting data. For example, railroads and the telegraph were integral to the development of NWS-COOP (as well as other observer networks) and the near-term weather forecasts it supported (Vetter 2011b). More recent advances in data management, online resources (e.g., Citizen-Science.org), and communications technology, as well as studies of the quality and value of citizen science data, have continued to transform the field (Dickinson and Bonney 2012, Bonney et al. 2014).

### Engagement and Science Literacy

Citizen science is also increasingly seen as a way to engage the public in science, improve scientific literacy and interest in science, and inform participants about particular topics, such as butterfly ecology, vernal pool conservation, or climate change (Lowman et al. 2009, Zoellick et al. 2012, Wals et al. 2014). This engagement is a major departure from most of the history of citizen science, when projects were set up mainly to achieve scientific objectives. Instead, many citizen science projects are now being organized in part as a means to improve participants' scientific literacy and understanding of the topics they are studying (Bonney et al. 2009). Scientists are also increasingly aware of the potential for combining historical citizen science datasets with current observations to gain insights into the ecological impacts of changes in climate, land use, and other drivers of environmental change. This renewed interest in citizen science, enriched with the perspectives and data provided by the long tradition of public participation in science, has the opportunity to broaden the engagement of the public in ecological research and lead to improvements in scientific education and insights.

## RECOMMENDATIONS FOR THE FUTURE

### Improving Our Understanding

Surprisingly little has been written about the history of citizen science, leaving much room for future research on the topic. Two recent, related projects, Constructing Scientific Communities (https://conscicom.web.ox.ac.uk) and Science Gossip (http://sciencegossip.org), are helping to fill this gap by digitizing historical scientific papers and documents and asking volunteers to identify the authors, the subject matter, and any species they portray. These documents will yield insights into important times in the history of citizen science and the whole of science.

## Using Historical Citizen Science

Rapid, anthropogenic changes in the environment currently have attracted a great deal of attention from scientists and government agencies. This attention is well deserved as climate change, ocean acidification, invasive species, land-use change, and other environmental perturbations are dramatically affecting ecosystems and forcing conservation organizations to rethink their approaches. Historical citizen science work provides a large source of data to inform the study of ecological responses to human actions and forecasts of future ecological responses. Some of the historical datasets that have been published in recent years, collected by both professionals and amateurs, have led to high-profile insights critical to our current understanding of ecological responses to anthropogenic change (Moritz et al. 2008, Willis et al. 2008, McClenachan 2009, Tingley et al. 2009, Burkle et al. 2013). Many other historical citizen science datasets have provided insights relevant to local ecology and conservation (Greene et al. 2005). Still other, similarly valuable citizen science datasets are likely waiting to be discovered in museums, libraries, government archives, and private homes (Hampton et al. 2013). We strongly encourage interested organizations to search for these caches of historical data and, when they have been found, to digitize and make them accessible.

Efforts to make data accessible are indeed occurring on a variety of scales—at individual museums and libraries, and across networks. For example, the National Park Service and the Schoodic Institute are working together to digitize the historical natural history collections held at Acadia National Park, which include many natural history datasets, journals, images, and biological specimens. Early analysis of these records and subsequent resurveys of plants and insects have demonstrated that the main section of the national park has lost 18 percent of its native flora over the past 120 years, while gaining hundreds of new species of insects (Greene et al. 2005, Chandler et al. 2012). Another example is the iDigBio project (www.idigbio.org), a large-scale effort to digitize biological museum specimens, many of which were collected by volunteers. As these historical citizen science data become accessible, more studies will be able to use them to explore changes in phenology, species distribution and abundance, and species interactions (Moritz et al. 2008, Bartomeus et al. 2013, Burkle et al. 2013, Scheper et al. 2014).

## ACKNOWLEDGMENTS

This chapter is adapted and updated from Miller-Rushing, Primack, and Bonney. 2012. The history of public participation in ecological research. Frontiers in Ecology and the Environment 10: 285–290. The findings and conclusions in this report are those of the authors and do not necessarily represent the views of the U.S. Department of Interior or the U.S. Government.

THREE

# Current Approaches to Citizen Science

FINN DANIELSEN, TEIS ADRIAN, PER MOESTRUP JENSEN,
JESUS MUÑOZ, JENNIFER L. SHIRK, AND NEIL D. BURGESS

The citizen science approach one chooses will depend on the context of the project. There are many different ways to do citizen science, and the choices made when designing a project will influence its outcomes (Shirk et al. 2012). Common to all citizen science projects is the involvement of participants. However, the amount and type of participation differ substantially from one project to the next. As a result, citizen science projects can be divided into five models based on the degree of participation (Shirk et al. 2012):

1. *Contractual projects,* where communities ask professional researchers to conduct a specific scientific investigation and report on the results;

2. *Contributory projects,* which are generally designed by scientists and for which members of the public primarily contribute data;

3. *Collaborative projects,* which are generally designed by scientists and for which members of the public contribute data but also help to refine project design, analyze data, or disseminate findings;

4. *Co-created projects,* which are designed by scientists and members of the public working together and for which at least some of the public participants are actively involved in most or all aspects of the research process; and

5. *Collegial contributions,* where non-credentialed individuals conduct research independently with varying degrees of expected recognition by institutionalized science or professionals.

Among these five models of participation, it is important to understand how the contributory, collaborative, and co-created approaches differ. In contributory projects, participants are involved primarily as data collectors, whereas in the collaborative and co-created approaches participants are involved in additional stages of the scientific process, including identifying the question of interest, designing methodologies, and analyzing data.

Citizen science projects can also be categorized by the types of activities in which the participants are involved (Bonney et al. 2016). We will discuss three such categories here:

Citizen scientists as data collectors

Citizen scientists as data interpreters

Citizen scientists as full partners

## CITIZEN SCIENTISTS AS DATA COLLECTORS

*How Does It Work?*

Projects in this category involve participants only in data collection. Design, analysis, and interpretation of the results are undertaken by professional scientists.

*Examples* In developed countries, participants are often volunteers who donate their time (e.g., to survey water and air quality, vegetation, weather, or populations of birds, amphibians, fishes, invertebrates, and invasive species; see, e.g., Støttrup et al. 2018). Commercially exploited wildlife populations are also surveyed by volunteers through such approaches as fisheries statistics, hunter records, and angler-diary programs (Venturelli et al. 2017). These citizen science projects often involve hundreds or thousands of participants whose efforts are embedded within a strong organizational infrastructure that provides sophisticated professional support and feedback to the participating volunteers. In developing countries, there are fewer examples of volunteer-based surveys (see chapter 17) and participants are more commonly paid to collect data as rangers working in protected areas, as staff on scientific expeditions, as staff assisting tourist volunteers doing survey work, or within hunter or fisher survey programs (Brofeldt et al. 2014, Chandler et al. 2017).

*Pros and Cons* In this approach, participants collect large amounts of data that otherwise would be extremely costly to gather. The skills required of participants are limited, the investment in training is small, and the interactions between professional scientists and participants are minimal. Sometimes the reliability of data collected by citizen scientists is questioned. However, the results of multiple studies demonstrate that such data are just as accurate and precise as data collected by professional scientists (Danielsen et al. 2014a; see chapter 9). Citizen science projects in this category are mainly of the contributory approach.

## CITIZEN SCIENTISTS AS DATA INTERPRETERS
### How Does It Work?
Projects in this category involve volunteers in data interpretation only. Professional scientists design the survey, collect the data, and analyze the results.

*Examples* In this category, we find citizen science projects with very large datasets that do not require a high degree of technical skill to interpret, such as images taken by trail cameras. Participants observe photos or videos and detect and classify specific, easily recorded features. Each classification is conducted by multiple participants, and the results are cross-validated. Examples of such projects include Camera CATalogue, Snapshot Serengeti, Snapshot Wisconsin, Western Shield Camera Watch, and WildCam Gorongosa. There are also examples of projects in which participants identify individual age classes of wildlife, such as adults, chicks, and eggs of penguins (PenguinWatch), or classify submerged kelp forests in satellite images (Floating Forest) or plankton in underwater images (Plankton Portal). Sometimes volunteers classify the behavior of wildlife in video recordings (Arizona BatWatch, Chimp&See) or hand-drawn pencil lines representing African rainforest trees' life-cycle events (Jungle Rhythms). There is even a project in which volunteers classify the similarity of spatial patterns within a river catchment, helping scientists model the hydrology of a river basin (Pattern Perception).

*Pros and Cons* The advantage of this approach is that it significantly reduces the time needed to interpret huge datasets from passive recording devices, which would otherwise need to be done by professional scientists. A potential challenge is inaccuracy of interpretation by the participants, though this is easily overcome by ensuring that the same images are interpreted by multiple people. Moreover, the basic knowledge required for interpretation can be provided by introductory training. This category is useful in surveys with large datasets in which the data's interpretation does not require technical skills but cannot be conducted by machines. It is particularly effective when patterns or features need to be recognized in many images. Projects in this category are mainly examples of the contributory approach.

## CITIZEN SCIENTISTS AS FULL PARTNERS
### How Does It Work?
Projects in this category involve citizen scientists in the entire research process—from formulation of questions and project design to data collection, analysis, and finally use of data in natural resource management, although professional scientists may provide advice and training.

*Examples* Projects in this category are often undertaken in areas where community members have some degree of control over the management of land and resources (Danielsen et al. 2014b). They are more participatory in character (Pocock et al. 2018) and are typically developed as part of an adaptive management plan. Most citizen science programs in the tropics and the Arctic belong to this category (Johnson et al. 2016). Sometimes they involve local and traditional knowledge held by communities who have long-term affiliations with specific landscapes (Zhao et al. 2016, Mustonen and Tossavainen 2018, Tengö et al. 2017). Examples in developed countries include volunteer wardens at nature reserves collecting data on which to base local management decisions and providing those data to national programs for larger-scale analyses. In the United States, projects of this type are often seen in relationship to environmental justice, where communities take up science as a tool to help address critical problems related to water, air, food, or personal health. An example is ALLARM (Alliance for Aquatic Resource Monitoring), a project in Pennsylvania that assists communities in addressing water quality concerns through data. Assistance is provided at all stages, from establishing protocols to interpreting data (Wilderman et al. 2004, Shirk et al. 2012). Examples from developing countries include community-based observation schemes— particularly those operating in community-managed protected areas, for instance in Namibia and other African countries (Danielsen et al. 2005, Chandler et al. 2017).

*Pros and Cons* While this category of project requires a large effort by the participants, it is also potentially very rewarding and beneficial to those involved. Benefits include participants having their voices heard, influencing how an area is managed, and contributing to capacity building and self-empowerment (Funder et al. 2013). However, if this category of project is to be successful, scientists must be able to facilitate a constructive dialogue with the participants. The category is particularly useful in areas where community members are closely connected to wildlife and the environment and where the government has a policy of involving and listening to community members in decisions on resource management (Danielsen et al. 2020). When digital platforms are used for storing and sharing data (Johnson et al. 2018), it may often be possible to connect and cross-weave with scientist-executed projects (Fidel et al. 2017). Projects of this category are mainly collaborative or co-created approaches.

## SUMMARY

Citizen science projects are often categorized by the degree of participation of volunteers, but they can also be categorized by the types of activities in which volunteers are involved. The first such category involves citizen scientists in data collection only and is useful in ecology and natural resource projects where large amounts of data need to be collected, an effort that would not be possible without a large number of participants. Even though citizen scientists only collect data, they are critical for the entire process, since there will be no data without their involvement. The second category involves citizen scientists only in interpretation of data and is useful when there is a very large number of items to be classified or interpreted (e.g., photos or other forms of data from passive recording devices) and when volunteer participation can significantly reduce the time spent by professional researchers in interpreting huge datasets. The last category involves citizen scientists in the entire scientific process, from formulation of questions to use of the data for conservation and management. This last approach is more demanding in terms of time and effort on the part of the participants, but the potential benefits are huge. Specifically, the full-participation approach can provide valuable data and, at the same time, help generate transparency, accountability, and local ownership in conservation and management initiatives, thereby empowering participants and prompting locally meaningful conservation actions. Since volunteers may be involved in a wide array of activities, their knowledge can play a greater role. While we recommend that you begin by thinking of particular types of citizen science projects, the reality is that each project is tailor-made to the particular needs being addressed and to the available resources and participants. Furthermore, regardless of the approach used, the amount of investment you put into the project is likely to influence the social cohesion and interactions among participants, and hence the quality of the work performed.

## ACKNOWLEDGMENTS

Finn Danielsen would like to thank the EC H2020 projects INTAROS (grant 727890) and CAPARDUS (grant 869673).

# PART TWO

# PLANNING AND IMPLEMENTATION OF CITIZEN SCIENCE PROJECTS

# FOUR

# Project Planning and Design

JOHN C. TWEDDLE, HELEN E. ROY, LUCY D. ROBINSON,
and MICHAEL J. O. POCOCK

If one is considering a citizen science approach to a project, effective planning and design are the keys to achieving and maximizing the benefits (Tweddle et al. 2012, Pocock et al. 2014). Getting the planning and design right will increase the chances of meeting your project objectives, be they scientific or focused on policy, conservation, learning, or participant experience. Citizen science is a highly flexible approach that can be used in a multitude of settings in terms of scale and methods, each with its own individual styles and goals (as reviewed in Roy et al. 2012, Pocock et al. 2018; also see chapter 3). Fortunately, a set of common principles of project planning and design can be applied in a wide range of settings, and we encourage the reader to refer to the Ten Principles of Citizen Science (Robinson et al. 2018) for additional guidance on good practice.

This chapter aims to provide guidance for the citizen science community by providing an introduction to the key topics to consider when designing a new citizen science project or updating an existing initiative. We assume that you have already identified the scientific question or hypothesis that you wish to study and are considering using a citizen science approach. Although our focus is on ecological or conservation-based projects, the guidance given here is broadly relevant to other disciplines. Similarly, while primarily focused on field-based projects, such as wildlife observation and environmental monitoring, many of the themes are transferable to online crowdsourcing projects such as those that invite participants to classify or analyze existing data (e.g., photographs or acoustic recordings; see www.citizensciencealliance.org). The ideas and recommendations presented here draw on key findings from a growing number of excellent, freely available resources that are helping to make citizen science an accessible and rewarding field of practice.

No two projects are the same, and there are many routes to success, along with numerous potential barriers and pitfalls. Hence, the ideas presented within this chapter are a starting point that you can add to and adapt to best fit your specific needs.

## MODELS FOR EFFECTIVE PROJECT DESIGN

A helpful way to consider a citizen science project is as a series of interconnected steps, each with their own goals and deadlines. These steps make a project seem less daunting while also identifying critical phases, timings, and resource requirements. In fact, many of the steps that apply to standard scientific research projects also hold for citizen science (Shirk et al. 2012):

1. Choose or define topic(s) for study.
2. Gather information and resources.
3. Develop specific questions (hypotheses).
4. Design data-collection methods (both experimental and observational).
5. Collect samples and/or record data.
6. Analyze samples and/or data.
7. Interpret data and draw conclusions.
8. Disseminate conclusions/translate results into action.
9. Discuss results and ask new questions.

Project design needs to ensure that overall research goals can be met, that sufficient data of appropriate quality can be gathered and/or analyzed, and that findings are reported through the most appropriate mechanism(s). *Where citizen science differs is through the involvement of the citizen scientists themselves, and this brings an additional, and exciting, dimension to project design.*

A range of strategies for developing a citizen science project have been proposed (e.g., Prysby and Super 2007, Bonney et al. 2009, Tweddle et al. 2012, Pocock et al. 2018). The features common to these strategies are (1) a structured approach, (2) an emphasis on pilot-testing the project with a subset of participants, (3) a willingness to adapt in response to feedback received, and (4) effective and timely communication with participants at all stages of the project. Many potential problems can be avoided by regular and thoughtful evaluation throughout the project's development (RCUK 2011; also see chapter 12).

Our planning and design approach is based on a well-established framework (Tweddle et al. 2012, Pocock et al. 2014) that provides a basis for introducing and discussing the key steps in planning and developing a citizen science project, many of which are discussed in further detail in subsequent chapters. This framework is based on a review of over 230 citizen science projects (Roy et al. 2012, Pocock et al. 2017) and considers the development and delivery of a typical citizen science project as five linked phases of activity (figure 4.1). The key steps hold for the majority of citizen science initiatives, irrespective of whether they are developed

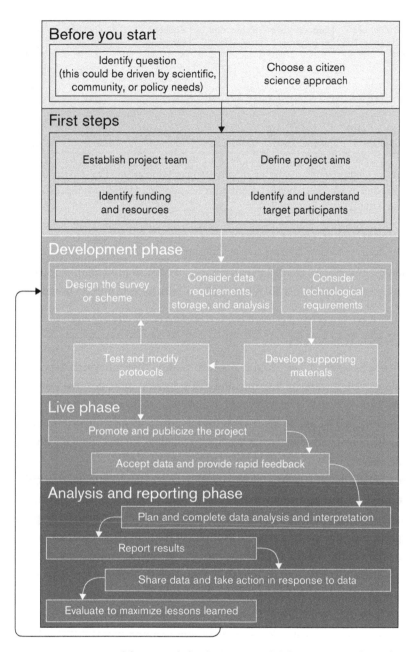

FIGURE 4.1 A proposed framework for designing and delivering an ecological or environmental citizen science project. Credit: designed by Heather Lowther, Centre for Ecology and Hydrology; reproduced, with permission, from Tweddle et al. (2012).

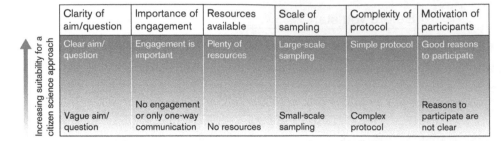

| | Clarity of aim/question | Importance of engagement | Resources available | Scale of sampling | Complexity of protocol | Motivation of participants |
|---|---|---|---|---|---|---|
| *Increasing suitability for a citizen science approach* | Clear aim/question | Engagement is important | Plenty of resources | Large-scale sampling | Simple protocol | Good reasons to participate |
| | Vague aim/question | No engagement or only one-way communication | No resources | Small-scale sampling | Complex protocol | Reasons to participate are not clear |

FIGURE 4.2. Is your project suitable for citizen science? Here are six broad areas to consider. Credit: designed by Heather Lowther, Centre for Ecology and Hydrology; reproduced, with permission, from Pocock et al. (2014).

by an academic institution, created in partnership between a professional organization and a community group to research a shared interest, or entirely community based and led.

## BEFORE YOU START

Citizen science has emerged as a highly popular method through which to generate new scientific understanding, support learning and skills development, address local environmental concerns, and/or foster ownership of local places. As an approach, citizen science can facilitate like-minded people to share their enthusiasm and knowledge and to develop new skills through shared endeavor. However, citizen science as a methodology does not work in all circumstances.

Before starting out, it is worth taking a step back and assessing whether citizen science is the best approach through which to meet the project goals (e.g., data collection, education, community building) that you have identified. Consider how involving participants will benefit your project and the participants themselves. Is participation essential and desirable, or will it detract from the overall aims of the project? Will your project appeal to participants, and be a practical and safe way to generate or analyze the volume and quality of scientific data that are needed? Can the project aims be clearly communicated to participants and are you willing and able to support their involvement by providing coordination, training, resources, and feedback? There are a range of other factors that you may wish to consider before committing to a citizen science approach (figure 4.2).

## FIRST STEPS

Assuming that a citizen science approach will be the best way to meet your research and other goals, the next step is to develop your project! This section outlines a few key ideas that will help you establish a stable foundation.

*Key Tasks*

Establish a project team.

Identify funding and resources.

Identify and understand potential participants.

Define and agree on project goals and objectives.

Consider ethical and legal requirements.

## Establish a Project Team

Whether a citizen science project is led by a researcher, a natural history group, an educator, a charity, or a community group, an effective team will be central to its success. A strong and enthusiastic team with relevant expertise will make for efficient and enjoyable project design and delivery. Irrespective of the style or type of project, you will need to access a variety of skill sets, from scientific skills to evaluation, communication, and marketing. Identify the specific skills required and aim to build a team with the appropriate experience. Each team member brings their own perspective and expertise and should use this in a positive way. Clarity of direction, identification of shared goals, and an environment in which all team members are listened to and have the opportunity to shape the project are keys to successful team dynamics. Notably, for co-created projects (see chapter 3) it is important that a team contains an equal balance of representatives from each of the communities that will be co-developing the project. Existing networks of naturalists and active community environmental groups can be good places to start.

Effective teams share ideas and expertise and can be an excellent way to pool resources, maximize publicity, and avoid duplication of effort. Include all interested parties in conversations, use accessible language, and be open to the views, ideas, and priorities of others. Such openness can be challenging if you are working with a broad range of stakeholder groups (e.g., Ramirez-Andreotta et al. 2014) or across academic disciplines. Thus, at the outset of a project, differences in language used, values, and priorities can be challenging, particularly for topics like environmental protection and nature conservation. However, with time and commitment from all parties, long-lasting trust-based partnerships can be built, to mutual benefit. Chapter 6 explores team development in more depth.

## Identify Funding and Resources

A common misconception is that citizen science is a free or low-cost option for undertaking science. When paired with appropriate goals, citizen science can certainly offer a cost-effective approach (e.g., Roy et al. 2012, Pocock et al. 2014, Blaney et al. 2016), but it is important to recognize that time and financial resources are required throughout the life span of the project. Thus, it is critical to consider the likely resource requirements early on (e.g., James 2011a) and compare these with the resources currently available, financial or otherwise. As mentioned above, working in a partnership can be a great way to pool funding, staff, and volunteer time, and to access in-kind contributions and physical resources. Blaney et al. (2016) provide an excellent interactive tool for cost-benefit analysis.

The total amount of resources required to plan, design, and deliver a project will be highly dependent on the nature of the project itself. However, you can consider three main categories of resources in your planning: staff or participant time, physical resources (e.g., equipment), and financial resources. The contributions of each of these three will vary throughout the project's life span, such that advanced planning is key to ensuring that appropriate forms and levels of resources are available at the right times. The early stages of project development can be particularly resource intensive in terms of staff or participant time, physical and financial resources (e.g., to test and develop protocols), and the design and production of project materials (e.g., website, participants pack, training resources). Remember that a project organizer's time will be required throughout to promote the project, support participants, and analyze and share results (Pocock et al. 2014). As such, your commitment and enthusiasm are vital to the success of the project.

If you decide to seek external funding to help deliver your project (e.g., grants, commercial sponsorship, or crowdfunding), try to plan as far in advance as possible and allow enough time to identify suitable funders and to complete the application process. Furthermore, it is a good idea to check funder priorities and expectations at the outset, as these will highlight what funders are interested in supporting and any requirements that will need to be met alongside your own project goals. If additional funding is critical to your project, but you do not have grant-writing or fund-raising experience at your disposal, this could be a great opportunity to partner with an individual or organization that can help.

## Identify and Understand Potential Participants

Think carefully about who the most appropriate audience(s) will be for your project and why. It can be appealing to try to develop a citizen science project for "anyone and everyone," but we are all different, with varying motivations, interests, skills, and concerns. Understanding the interests, abilities, and expectations of potential participants is key to both encouraging their involvement and designing effective project materials, because people participate in outdoor ecological citizen science projects for a wide range of reasons (see table 4.1). Similarly, people take part in online crowdsourced projects for various reasons, including contribution to research efforts, interest and enthusiasm for the subject, personal learning, and cooperative endeavor.

In project design terms, what appeals to and works for one group of potential participants (e.g., naturalists) may be less effective with another (e.g., schoolchildren). Focusing on a clearly defined type of participant is an effective approach. Experimental protocols, technologies, training, reporting, and even the style of language employed will be most effective if tailored to the specific audience or audiences that you are inviting to participate. Talk to potential participants as early as possible. What are participant interests and motivations, and can your project build on these? How would they like to contribute their time and skills, and do they have specific technology, access, or training requirements? What do they perceive as barriers to participation?

TABLE 4.1 Examples of Motivational Factors for Involvement in Ecological Citizen Science Projects by Different Stakeholder Groups

| Amateur naturalists | Local communities | Educators |
|---|---|---|
| Interest in project | Interest in project | Interest in project |
| Contributing to a shared scientific endeavor that matches personal conservation and environmental values | Contributing to a shared scientific endeavor that matches personal conservation and environmental values | Contributing to a shared scientific endeavor that matches personal conservation and environmental values |
| Enjoyment of finding, identifying, and recording wildlife | Deep affinity with location (historical, social, cultural value) | Participation in "real" science |
| Aesthetic appreciation of nature | Projects that support shared, locally relevant goals | Fit with curriculum |
| Rapidity of feedback | Rapidity of feedback | Rapidity of feedback |
| Personal learning and reputation | Opportunity to share knowledge | Personal confidence in subject area (e.g., ability to identify the species concerned) |
| To support conservation activity | New activities in a familiar space | Ease of participation; ready-made, accessible materials |
| Social aspects | Social aspects | Location and cost |
| To spend time outdoors | To spend time outdoors | Teaching outside the classroom |

Sources: Ellis and Waterton 2005, Grove-White et al. 2007, Hobbs and White 2012, Roy et al. 2012, Tweddle et al. 2012, Robinson et al. 2013, Geoghegan et al. 2016.

As humans, we often make assumptions concerning what will appeal to others or work based on our own personal experiences, and these are unlikely to be representative. Much in the same way that scientists collect pilot data, one of the best ways to learn about participants is to run a small-scale trial. Make contact with your target audience, present your ideas, and give them a chance to try out and comment on your initial protocols. Reflect on the experience, evaluate, and be prepared to alter the approach—or even completely redesign the project. Chapter 7 further explores methods for maintaining the support of citizen scientists in the context of longer-term projects.

Consider accessibility, equity, and integrity from the outset. For example, it is quite possible that decisions made in good faith at the start of the project may unintentionally result in the exclusion of certain demographics and communities. If your project specifically aims to increase participation from an underrepresented group or community, then time spent getting to know the values, interests, and potential barriers to participation within this community is essential. Visit the community at a venue of their choosing, be open about your goals and accepting of differing views and values, and ensure, above all, that you make time to develop the partnership. If you are seeking to engage with communities that are traditionally underrepresented in science or the environment, then you may want to conduct some additional reading specifically focused on this topic (Pandya 2012).

### Define and Agree on Project Goals and Objectives

Once you have identified available resources and the expertise and interests of the team and potential participants, it is valuable to refine your goals, as what was initially considered and what you have in place may differ. Citizen science projects often have multiple aims, from undertaking novel research, to addressing policy needs, to engaging participants with local environmental or conservation issues. In addition, you may want to include specific goals that focus on the participants themselves, such as goals related to education and skill development, attitudinal or behavioral change, or capacity building within a particular community.

Managing multiple goals (e.g., engagement, education, and data collection) can be demanding. As a result, it is easy to lose sight of some goals and allow one to dominate. Thus, be clear about what balance you are trying to achieve. Team members may have differing priorities, so ensure that communication is open and effective. Aim to establish consensus and strike a balance that everyone is happy with. As with any project, it is important to agree on goals at the outset and to establish processes that provide opportunities for evaluation and review. Agree on an end point, on what success looks like, and, importantly, on how it will be assessed. We stress that engagement and dialogue about science is a worthwhile activity for its own sake (see Bowater and Yeoman 2012), but engagement alone is not citizen science. If you would like to learn more about how scientific and social outcomes can work together to influence the success of nature conservation activity, Kapos et al. (2008), McKinley et al. (2017), and Ballard et al. (2017) provide helpful summaries.

### Consider Ethical and Legal Requirements

Once you have identified your project goals and objectives—and before you start to develop the project in earnest—we recommend that you confirm that the project will run according to best practices. When running a citizen science project, you have a degree of duty to care for the participants. Your project needs to meet ethical guidelines (see chapter 5; Resnik et al. 2015, Robinson et al. 2018) and other relevant policy and legal considerations, including public liability, risk assessment, intellectual property rights (Scassa and Chung 2015), data protection and privacy (especially if collecting digital information about participants; see Bowser et al. 2013), and any applicable local laws (such as land access). Data protection needs to consider both your storage and use of the data and any third-party accessibility to the data, including passing on to repositories for wider sharing. Chapter 5 explains the policy and legal aspects of citizen science in more detail.

At first glance, the first steps of a project may look daunting. But the public nature of citizen science means that it is important to consider these first steps thoroughly because getting them right will protect you as a project organizer, in addition to protecting other stakeholders and, most importantly, the participants themselves.

## DEVELOPMENT PHASE

The development phase of the project focuses on design and testing of protocols, tools, and resources that will be used. Remember to regularly check progress against aims and objectives.

*Key Tasks*

Research existing projects to avoid duplication.

Design the survey or scheme.

Plan how you will evaluate success.

Consider data requirements, storage, and analysis.

Consider technological requirements.

Develop training and supporting materials.

Test and modify protocols.

### Research Existing Projects to Avoid Duplication

Search for other, similar projects and try not to reinvent the wheel. Most practitioners are keen to share ideas and experiences and will be open to collaboration. Look for opportunities that will benefit both you and your potential collaborators and consider whether your protocol will produce or analyze data that can be shared or added to existing datasets to provide greater value.

### Design the Survey or Scheme

Design a protocol that will enable participating citizen scientists to efficiently gather and/or analyze the data appropriate to your scientific aims. In general, there will be greater participation in (1) simple projects versus those perceived to have complex and demanding protocols, (2) short-term versus long-term projects, and (3) single-visit versus multiple-visit projects. However, it may be appropriate to develop a complex protocol with the expectation that a few highly skilled or motivated participants will be involved from the outset, or that participants may progress from a simple to a complex protocol as they develop skills and deepen their engagement. Indeed, some citizen scientists will thrive on the complexity. Notably, there is often a trade-off between the complexity or length of a task and the number of participants, and this can be challenging to manage. It is generally good practice to avoid unnecessary complexity, while still enabling capture of the data you require (that is, make it "as simple as possible but no simpler"). Finally, ensure that protocols are appropriate for the target audience and aim to minimize the number of steps between data collection and data submission.

For field-based citizen science projects, consider the types and volumes of data that you need to collect and the spatial and temporal resolution required. Is it feasible to standardize the methods so that all participants gather or analyze data in the same way? Such standardization can greatly increase the range of statistical analyses that are possible. Additionally, can participants choose their own survey sites, or will they need to be pre-allocated? How will participants send their results to you—via an online data-entry form or smartphone app, or will they need to submit paper-based results or physical samples by post? If any equipment is required, consider how it might be obtained by participants.

The health and safety of participants is paramount, so test your protocol under realistic conditions and ensure that you write a risk assessment and standard operating procedures (note that these are required by many organizations). Amend the protocol to minimize any identified risks and give guidance where necessary (e.g., for the safe handling of equipment, wildlife, or samples). Similarly, consider any unintended harm that could be caused to people or the environment by your proposed activity and check any relevant legislation.

Keep in mind the expectations and skills of the target participants throughout the project. Is the activity appealing and realistically achievable? You may wish to develop a tiered protocol that offers simple and/or quick activities for beginners or participants with limited time, through to more complex activities for those who wish to progress or be more involved from the outset. Remember to keep in mind that the language and technology you employ can potentially exclude some individuals and communities. Understanding (and working with) your intended audience (see above) can help identify and then avoid this issue.

### Plan How You Will Evaluate Success

Evaluation is an ongoing process that can improve your project and help you measure the outcomes, in terms of both the data collected or analyzed and the experience that your participants receive. Evaluation (which is often done too late to be helpful) should be planned at an early stage and resourced throughout to be most effective. It is important to evaluate against the specific goals you set at the outset of the project. Consider working with an evaluator if your team does not have relevant experience.

Front-end (or baseline) evaluation occurs at the start of project development and provides a baseline understanding of an audience from which to measure change over the project (e.g., in their scientific knowledge, attitudes, and interests). Findings can be used to refine project aims and goals and help ensure that they fit with audience needs and interests. Formative (or implementation) evaluation is carried out as you develop the project and can help you determine strengths and weaknesses of the project's design and fine tune your protocol. Finally, summative (or impact/outcomes) evaluation happens at the end of the project or during a natural break. This summative evaluation can help you measure the effects and impacts that your project has on science, participants, policy, and so on (Tweddle et al. 2012, Phillips et al. 2014). Evaluation is discussed in detail in chapter 12.

### Consider Data Requirements, Storage, and Analysis

Plan how you are going to manage the data that are received, store them securely, and make them available in the long term (see chapter 10; also see James 2011b, Wiggins et al. 2013). As a general rule, try to maximize the value of your data to others by providing open access to nonsensitive data and by using accepted data and metadata standards and licensing.

These agreed-upon formats for storing and describing your data will make it easier for you to share your information with others (see chapter 10).

From the outset of your project, it is important to be transparent regarding how data will be stored and shared, particularly where there are data protection, data conservation, policy, or legal implications. This transparency will give members of the project team, potential participants, and other stakeholders the opportunity to raise concerns before issues arise. We strongly recommend that you develop data policies appropriate to your project (Bowser et al. 2013; also see chapter 5).

As with any scientific research, it is critical to plan a data-analysis approach before beginning to collect data, as this will help confirm that your protocol will generate the required information. Think carefully about the scientific question that you are investigating, or the hypothesis that you are aiming to test, and create a detailed breakdown of the data that you need to gather and analyze. Double-check who is best placed to complete the constituent tasks, which tasks can be best delivered by participants, which should be completed by other members of the team, and when you may need to recruit a specialist such as a statistician. You should also consider possible sources of error and bias in order to design a protocol and analyses that minimize and/or quantify their influence. Finally, double-check that you have incorporated data quality controls into your protocols.

The quality of data collected by participants is heavily influenced by the survey design, training materials, and support you provide. As with any project, it is important both to minimize the opportunities for errors and to understand how data quality varies between samples or even participants. Data of "known quality" (i.e., when accuracy and precision are quantified, even if not necessarily perfect) are scientifically useful and are also more likely to be used as evidence by policymakers, whereas data of unknown quality are open to scientific criticism. One of the most cost-effective ways to maximize data quality is by testing and retesting project protocols with potential participants. As well as helping to optimize the protocol itself and quantify data accuracy, testing can highlight commonly made mistakes and sources of bias and inform project training.

Data also need appropriate verification. Verification could include assessing a subset of data, looking at submitted photographs, or confirming that grid references match place-names. Online quizzes or self-assessment tools can help, but these will be inappropriate for some projects and, if not presented with care, can sometimes discourage participation (not everyone likes to feel they are being tested). Data quality and analysis are considered further in chapter 9.

## Consider Technological Requirements

Technological developments have revolutionized citizen science. Web-based data capture, analysis, and presentation tools and smartphone apps are in common use and are increasingly easy to develop (see Roy et al. 2012, August et al. 2015). A good website or app can enable rapid data entry, data validation, and real-time mapping of participants' results. Furthermore, a website or app can host background materials, training resources, and forums and blogs through which participants and project developers can interact as part of a project community. A wide range of cheap (including DIY), user-friendly environmental sensors are increasingly available, and field-based molecular biology tools will soon become an affordable possibility.

While technology is bringing exciting innovations to citizen science and can capture people's imaginations, it is worth remembering that technology can also form a barrier to participation (Tweddle et al. 2012) and that your choice of technology will have a direct influence on who takes part. If you opt for a high-tech approach, consider whether you can also accept data through more traditional routes (e.g., paper-based forms) to make the project accessible and appealing to all. Be aware that technologies develop and move on quickly, and an apparently cutting-edge technology can rapidly become out-of-date. Of course, cutting-edge technology can also rapidly become mainstream (e.g., the adoption of smartphones), so there is value, but also a high development cost, in adopting new technologies. All technology has a resource implication, so focus on employing technology that is appropriate and appealing to your audience and make sure it adds value to the project.

### Develop Training and Supporting Materials

A common reason for participants not submitting their data is a lack of confidence in the accuracy of their results. Well-supported citizen scientists will not only produce higher-quality data, but they will gain far more from participation (e.g., increased skills, improved confidence, new social networks). Consider the types of supporting materials that participants will need and the most appropriate forms they might take. These materials could include printed or online instructions, identification guides, lesson plans for schools, and any physical equipment that is needed. Video clips can be an excellent way to demonstrate techniques and introduce your team. You may also wish to develop a Frequently Asked Questions page, or a web forum that allows participants to share skills and experiences.

Pick the format that works best for your participants, but, as a general rule, minimize the amount of text. User testing is vital, so develop these resources with input from potential participants. Try to reuse existing materials where possible. Many groups will be happy to share resources they have produced if they are acknowledged appropriately—better still, they may be interested in contributing to your project.

Face-to-face training can be an excellent way to build skills and retain involvement in a project, but it is resource intensive. Mentoring can and should be an option for longer-running projects. Training is covered in greater depth in chapter 8.

### Test and Modify Protocols

This is the final and, in many ways, most critical aspect of the development phase. Test the survey protocols, data-entry forms, any analytical tools, and training materials with participants, under realistic conditions. It can be tempting to skip testing when deadlines get tight, so allow plenty of time for this activity. Observe how people use materials, and solicit feedback. Where do participants go wrong and what questions do they ask? Consider all constructive feedback and adapt your protocol and supporting materials accordingly. How can you amend your approach to maximize the clarity of materials, quality of data generated, and user experience? Testing by users is often an iterative process, so be ready to retest and adapt again if necessary. Be prepared to make radical changes to your project if feedback from participants suggests that it will not be successful in its current form.

## LIVE PHASE

The live phase is when publicity efforts will peak and participants will be gathering or analyzing the data that the scientific aspects of your project depend on. The live phase can be the most enjoyable and rewarding part of the project, as you see enthusiasm growing and data or samples come flooding in. However, the live phase is also when unexpected glitches are most likely to appear, so be prepared to respond. Allow plenty of time across the project team to publicize the project, support participants, and fix any problems that arise.

*Key Tasks*

Promote and publicize your project.

Accept data and provide rapid feedback.

### Promote and Publicize Your Project

If it fits with your project goals, you may need to promote and publicize your project to generate participation. The level of effort needed for publicity will be highly dependent on the specifics of your project, but it is another aspect of project delivery that will benefit from time spent understanding the motivations and expectations of potential participants (Geoghegan et al. 2016). Publicity can be relatively straightforward for co-created and community-led projects, as the key stakeholders will already be actively involved and can help spread the word through their contact networks. Publicity may be most challenging for contributory projects that depend on attracting a large audience of citizen scientists over a dispersed geographic area. In such cases, develop a communications plan and be prepared to spend a lot of time and effort promoting your project if it is not part of an existing, well-known scheme or web platform. If you do not have experience with marketing or working with the media, consider partnering with someone who does. Remember that the methods you use to recruit participants will play a large role in determining who participates.

Manage expectations from the outset by being honest and not overstating the likely impacts of the project. Target your efforts on the publicity outlets that are most likely to reach your chosen audience(s). Do you need national publicity, or are you aiming at a more local audience, such as a place-based community group or natural history society? Existing networks can be an excellent place to start, especially if their interests and goals match those of your project.

Prepare in advance and tailor the messaging to the participant groups you are aiming at, using clear and accessible language that communicates your key messages as succinctly as possible. If you only had thirty seconds to explain your project to a potential participant or journalist, what would you say? Why do you need their input? What are you inviting people to do, and how can they get involved? Try to explain why your project is personally or socially relevant (i.e., its value) to those you are aiming to reach, as this will be key to gaining their support. The following list summarizes some of the publicity outlets that have proved successful for promoting citizen science projects:

Existing contact networks

Websites: project site, project partner sites, online databases of citizen science projects

Press, TV, and radio

Social media via blogs or platforms such as Facebook and Twitter

Face-to-face communication (e.g., talks, workshops, and events)

Launch event (e.g., a BioBlitz)

Email newsletters and newsgroups (e.g., for school teachers or special interest groups)

Posters and fliers in local libraries, community centers, or parks

Pick those that will be appropriate to your audience, be innovative, and be prepared to try more than one approach if necessary.

Negative messaging can put people off, so try to focus on the positive aims of the project (e.g., generating knowledge or conducting a conservation activity). Emphasizing the positive can be effective even when dealing with fundamentally challenging issues such as the impacts of environmental change or the spread of invasive species.

Finally, keep in mind that retention of participants is a critical aspect for projects that require ongoing contributions, such as community-based environmental monitoring initiatives. In some cases, people's motivations for initially getting involved in a project may differ from their motivations for staying involved once the initial task is completed (Geoghegan et al. 2016). Participant retention is discussed further in chapter 7.

### Accept Data and Provide Rapid Feedback

The procedures established during the development phase will handle the submission of data and/or samples. During and after submission of data and samples, we recommend providing feedback to the participants and, wherever possible, engaging in two-way dialogue or building a discussion forum around the project for the whole community. As a minimum, feedback should include thanking participants, which shows they are valued and can be a powerful motivator in encouraging continued involvement as well as providing a sense of purpose and achievement.

Routes for providing feedback depend on your budget and the mechanisms through which you will be receiving data and reporting results. When possible, rapid feedback is preferable. Initial feedback could be as simple as an automated email thanking participants for submitting their results. If you have time, or where it is desirable to establish a relationship with each participant, you may wish to consider thanking them personally. This personal touch is a great way to highlight an interesting find or support participants to further develop their skills, but be realistic about whether you have the resources to undertake and sustain this approach. For projects that work with a specific community, face-to-face contact is invaluable. Try to factor in as much time as possible to provide feedback and results, and discuss other aspects of the project, ideally at a venue within the community.

If you cannot thank participants individually, ensure that it is easy for them to both ascertain that their data have been received and find out how the project is progressing. For wildlife observations or environmental data, an up-to-date results map on the project website can be particularly effective. Seeing your records appear on a map not only shows that data have been received but can be a powerful motivator, particularly with family and school audiences.

Try to provide a regular project update that summarizes results and activity to date. Updates could be through the project website, a newsletter, a blog or email, or a short printed update on a community-center or park notice board. Social media can also provide an effective and rapid method for feedback, though its reach will be limited to followers in the platforms that you target.

## ANALYSIS AND REPORTING PHASE

For some projects, the analysis and reporting phase is the final stage, during which results are analyzed and communicated, data are shared and archived, and a final evaluation is completed. For long-term projects, analysis and reporting can present an opportunity to share the results that have been obtained to date and to improve project protocols.

*Key Tasks*

Plan and complete data analysis and interpretation.

Report results.

Share data.

Take action in response to data and evaluate success.

### Plan and Complete Data Analysis and Interpretation

Data analysis and interpretation are time-consuming elements of any scientific project, so plan accordingly. Moreover, as with any scientific dataset, it is important to understand sources of error or bias and to account for them when interpreting results. One way to increase data quality and reduce time on data cleaning is through the inclusion of validation and verification processes in project design (see chapter 12). Finally, which statistical analyses you choose for your data will depend entirely on the type of data collected. If in doubt, enlist the help of a statistician (though ideally you would have already done that in the project design phase).

Alongside the scientific data, you may also need to analyze qualitative or ethnographic data, particularly if one of your goals is to influence behavioral change or learning within your participants. Chapter 13 looks in depth at how this can be approached.

Whether your project contains societal or participant-focused goals or not, it is worth considering whether you wish to incorporate any qualitative data into your interpretation and project reporting (e.g., comments or feedback that you have gathered from participants). Qualitative data can be more labor intensive to evaluate, but they may provide valuable insights or useful evaluation and feedback for improving the project. Data quality and analysis are considered further in chapters 9 and 12.

## Report Results

Reporting the results of a project is a critical aspect of all scientific studies. For citizen science projects, reporting results is particularly important because stakeholders can be both varied and numerous: from the citizen scientists who have made the project possible, to project partners, data users, academia, conservation agencies, and the press. Communicating your findings ensures that the knowledge generated by your project is shared, and it rewards the participants and other stakeholders who have contributed their time and expertise. We all want to see that our efforts have been productive (Tweddle et al. 2012, Geoghegan et al. 2016).

Effective communication of your results will help maximize the utility of your data and findings, whether for practical conservation and environmental improvement or for academic or legislative purposes. Be mindful that it can often take a period of months to years to collate, analyze, and then publish the scientific findings, or to implement conservation measures that may be informed by your results. It is important that participants are kept up-to-date while the publication or communication process progresses.

Different audiences will require different key messages, levels of detail, and types of data visualization. Participants will be interested in seeing how their efforts have helped the project in terms of general trends, summary statistics, and visualizations of the data. Where possible, it can also be very effective to highlight the local relevance of results. You could also consider presenting interviews with participants or inviting them to contribute blog posts (for example) so that they can share their experience of the project. Scientists and policymakers will be interested in the broad results—what your data show and how the results fit with the wider picture—but also in the details (methods and analyses employed, data quality and confidence levels, and comparison with previous understanding). Media and press look for short, snappy phrases that explain what you did, for what purpose, and why your results are so fascinating. Make your communication relevant to the audience (viewers or readers) by phrasing it in a way that will capture their interest and imagination. Regional press will be particularly interested in local stories.

Aim to report your results by the most appropriate route for the audience, which may include some of the feedback mechanisms discussed above. Face-to-face presentations and question-and-answer sessions can have a fantastic impact and should form a key element of reporting for community-based projects. Always highlight the input of the citizen scientists.

Finally, failure to communicate results to stakeholders is sometimes cited as a stumbling block by project organizers. It can be tempting not to communicate results if a project did not go as planned. However, there is always something to report back to stakeholders, so do not be afraid to talk about the challenges you encountered, alongside the successes of the project. Not only does this openness convey an accurate representation of the scientific process to project contributors, but it also helps other practitioners. Chapter 11 explores the topic of reporting in more detail.

### Share Data

Sharing nonsensitive data as widely as possible and in a format that can be used by others will greatly increase the value of the project and, thus, that of the contribution the participants have made. Remember to consider potential intellectual property rights and data protection requirements at an early stage (chapter 5) and try to make data available in electronic format with a machine-readable license wherever possible, as this will increase their accessibility to others. Data can be hosted on your own website or in dedicated repositories. Ensuring that project data are stored in a secure and accessible place will reassure participants of the long-term value of their efforts. Finally, to maximize the value of the data, it is important to consider how data can be actively shared with others, not just passively made available. For example, wildlife records should be shared with repositories that are part of the Global Biodiversity Information Facility (GBIF). Further information on how best to store and share the data your project has generated is available in chapter 10.

### Take Action in Response to Data and Evaluate Success

So you have conducted your project, reported the findings to key stakeholders, and ensured that the data are archived and accessible to others. What happens next will be highly dependent on your project and resourcing going forward. Here are some things you might do:

- Close the project and celebrate its successes! For projects with websites, provide an archive of resources (which could be used for public engagement with science activities) and results and provide a clear summary of the main messages of the project. Consider the participants and direct them and new website visitors to other similar projects (run by you or by others).

- Continue to build on any partnerships or new ways of working that have been established.

- Pursue any new research questions that arise from the results (which could be through citizen science or other research approaches).

- Take action in response to the scientific (or social) findings. In some cases, appropriate action may arise from the information gathered—for example, the containment or removal of populations of invasive species that your project has identified. In other cases, citizen science can inform policy actions and management of specific sites.
- Complete the project evaluation (see above and chapter 12) so that you can learn from your experience. What worked? What would you do differently next time? Consider sharing your experiences with other practitioners.

### HOW TO FIND OUT MORE

We hope this chapter has given you a useful introduction to the main factors that need to be considered when developing a citizen science project. These factors will all be elucidated further in the following chapters. If you are interested in exploring these themes in more detail, please explore the cited references. Another fantastic way to understand particular issues is to get in touch with other practitioners and learn from their experiences.

# FIVE

# Legal, Ethical, and Policy Considerations

ANNE BOWSER, ANDREA WIGGINS, and ELIZABETH TYSON

## LAW, ETHICS, AND POLICY IN CITIZEN SCIENCE

If a participant is injured while contributing to a citizen science project, who is liable for the cost of their recovery? How should data-sharing ideals be balanced with proprietary uses, through publication for example? And how should data-sharing ideals be balanced with the need to protect volunteer privacy? This chapter addresses these questions and other legal, ethical, and policy issues.[1]

Law, ethics, and policy are three types of guidelines that describe and prescribe interactions between different social agents, including individuals, community groups, formal organizations, and political bodies. *Laws* are the most stable of these and are enforceable by government entities. For example, in the United States, the Children's Online Privacy Protection Act (COPPA) is enforced by the Federal Trade Commission (FTC). Parties that violate COPPA are subject to civil prosecution by the FTC and to fines.

*Ethics* are general moral guidelines and are the least stable of the three types of guidelines we discuss. For example, many ethics decisions are determined by the specifics or context of a particular situation. Asking volunteers to collect soil samples may be generally considered ethical if the samples are collected from some public properties, such as the yards of public schools, but not from others, such as protected national parks. Asking volunteers to

---

1. The authors are not legal experts, and this chapter is neither offered nor intended as legal advice. Only you know what your project does and what it needs. In addition, while this chapter offers examples primarily from the U.S. legal paradigm, similar legal, policy, and ethical constraints apply in other countries.

collect soil samples on private land may or may not be ethical, depending on how sensitive data, such as the full names and ages of participants who collected samples, are treated.

*Policy* refers to a collection of formal and informal principles that guide decision making and are developed iteratively over time. Policy is thus more flexible than law and less subjective than ethics. Some policy directly reflects law, while other policy reflects interpretations of the law, such as cases heard by the Supreme Court. Some public policies may become law, while other policies remain lesser guidelines. Policy is designed to be more flexible than law so that it can incorporate new precedent cases and research on a particular topic.

In the citizen science context, policies are collections of documents that dictate how one party, such as a citizen science project coordinator, may interact with another, such as a citizen science participant. Such policies are a crucial, but often neglected, feature of project design. They specify the legal and ethical responsibilities that projects have toward participants, such as protecting participant privacy or mitigating physical and emotional harm. Policies also clarify the responsibilities participants have toward projects, such as taking full responsibility for injuries gained in the course of service or promising to adhere to community standards of conduct. Finally, data policies clarify data ownership and management.

*Data policy* codifies and dictates the practices of organizations in relation to data (Bowser et al. 2013). Because citizen science activities that involve both professionals and volunteers are often related to data (collection, processing, etc.), data policies are more salient in citizen science than in traditional scientific research conducted from within organizations or other formalized partnerships. Common types of policies include the following:

> *Legal policies* address liability concerns and may describe compliance with existing law.
>
> *Privacy policies* deal with how data, including personally identifiable information, are collected and stored.
>
> *Terms of use* dictate how a product (including data) or a service (including data collection and/or analysis) can be accessed and used.

These policies become meaningful when embedded within a *user agreement* or a legal contract between a project and its volunteers.

Throughout this chapter, when we identify an important law, ethical principle, or public policy, we also offer guidelines for how citizen science projects can write their policies to demonstrate alignment. We cannot overemphasize the importance of having formal policies and a user agreement in place. In some cases, documentation of data practices is required as a component of legal compliance. Documentation may also be a prerequisite if practices are to be recognized by a court of law. That said, we understand that scientists and project coordinators are driven by their passion for research or education conducted with the help of citizen science volunteers. Thus, to make legal, ethical, and policy considerations more relevant, we use a model of the citizen science research process to frame the concerns

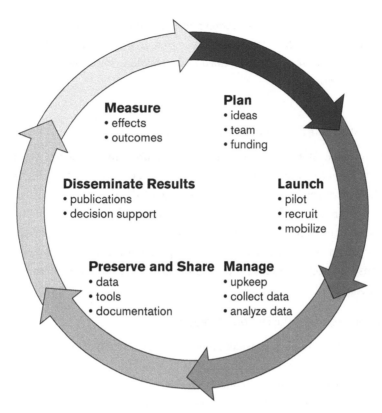

FIGURE 5.1. Research life cycle for citizen science. Key steps include *plan, launch, manage, preserve and share, disseminate results,* and *measure.*

facing practitioners. This chapter covers the six stages of the research life cycle (figure 5.1) and discusses key considerations at each step.

## THE RESEARCH LIFE CYCLE

Figure 5.1 shows the typical steps in managing citizen science projects from creation through preservation and discovery. The activities involved in this research life cycle are nonlinear. In other words, within an ongoing citizen science project, multiple project- and data-management processes may occur at the same time, occur in a different order than is shown in figure 5.1, or skip steps that are not applicable. At the same time, decisions related to every life-cycle stage are made during planning. One of the advantages of using a life-cycle model is that it readily supports long-term, big-picture planning that can be helpful when considering policy decisions. Notably, the research life cycle is similar to the project planning and design framework discussed in chapter 4.

Just as the research life cycle is cyclical rather than linear, data policies should be established during a project's planning stages and periodically revisited. So, while policies about sharing data are covered in the preservation and discovery sections of the research life cycle, these need to be in place before data collection begins. In the remainder of this chapter, each section of the research life cycle is described in depth, along with the primary legal, ethical, and policy considerations.

## RESEARCH LIFE CYCLE, PHASE 1: PLAN

The planning stage of the research life cycle includes mapping out the policies and processes for the entire life cycle, starting with project goals and working backwards to develop policies that meet the project's needs. One of the first steps in planning is determining the relevant types of legal, policy, and ethical guidelines that will affect a project. Some considerations—including compliance with certain laws—are unique to projects with specific characteristics (e.g., projects supported by U.S. federal agencies). Other considerations—like ethical deliberation and/or formal review—are common to all citizen science projects.

### General Legal and Policy Considerations

Some legal, ethical, and policy considerations are determined by funding source or institutional involvement. As a result, projects with shared funding sources and institutional requirements often express similar concerns (table 5.1).

Projects that *are run by* U.S. federal agencies, or projects that receive direct financial support from these agencies, express common legal concerns (Gellman 2015). For example, federally funded projects must comply with the Privacy Act of 1974, the Paperwork Reduction Act, and the Freedom of Information Act, in addition to ethical regulations. Citizen science projects that *receive grants from* U.S. federal agencies—such as the National Science Foundation (NSF), the National Institutes of Health (NIH), and others—are subject to specific guidelines as a condition of the award. For example, NSF and NIH grantees must obtain approval for work with human subjects from an institutional review board (IRB) and for animal use and welfare from an institutional animal care and use committee (IACUC). Human subjects review often applies to social survey instruments or projects involving data collection from or by people, whereas animal use oversight ensures the ethical treatment of both laboratory animals and wild vertebrates. Other project-specific guidelines—such as those dealing with age-based discrimination, which may impact some projects, and requirements for data sharing and management plans—are detailed in grant policies (e.g., the NSF's grant policy manual; National Science Foundation 2019).

Nonfederal funders, such as private foundations, may have different requirements. For example, the Alfred P. Sloan Foundation suggests that, in line with their core values, information products—including publications, code, and presentation materials—should be offered as open source. Citizen science projects should review the grant policy manuals of their individual funders and negotiate any differences between funder policies when support comes from multiple sources.

TABLE 5.1  Common Laws and Policies Affecting Citizen Science Projects

| Policy or law | Affected parties | Purpose | Alignment |
|---|---|---|---|
| Institutional review board (IRB) approval | Federal grantees; projects run by most federal agencies; projects run by many nonprofits and academics | To support ethical human subjects research | Apply for exemption or for approval to an internal IRB third-party service |
| Institutional animal care and use committee (IACUC) approval | Federal grantees; projects run by most federal agencies; projects run by academics | To support ethical research with vertebrates, including wildlife | Apply for approval |
| Paperwork Reduction Act (PRA) | Projects run by federal agencies | To reduce the public's information burden | Complete Office of Management & Budget review[a] |
| Freedom of Information Act (FOIA) | Projects run by federal agencies; federal grantees; state agencies (varies by state; may be under a different name) | To support transparency of government data collection | Avoid collecting personally identifiable information; eliminate ability to retrieve records by personally identifiable information; be prepared to share records upon request |
| Children's Online Privacy Protection Act (COPPA) | Projects run by private-sector companies or federal agencies that collect personally identifiable information from children under thirteen | To protect the privacy and safety of children under thirteen | Post a privacy notice and notify parents of data collection |
| Volunteer Protection Act of 1997 | Projects run by nonprofits; most projects run by academics; projects run by federal agencies | To encourage volunteerism by reducing organizational liability | Have volunteers sign a liability waiver |
| Antideficiency Act | Projects run by federal agencies | To prevent government agencies from taking unfair advantage of volunteers | Clearly define the scope of a citizen science activity; have volunteers sign a waiver |

TABLE 5.1  *Continued*

| Policy or law | Affected parties | Purpose | Alignment |
|---|---|---|---|
| Privacy Act of 1974 | Projects run by federal agencies that retrieve database records by a personal identifier | To protect the privacy of U.S. citizens | Retrieve database records without a personal identifier, or file a System of Records Notice (SORN) |
| Open data and open access policies | Projects generating peer-reviewed scientific publications at relevant institutions (open access) or via relevant venues (open data) | To support scientific integrity and access to data and results | Familiarity with requirements of local institutions regarding open access to research results, and publisher requirements regarding open data; these organizational policies can affect data access decisions |
| Land ownership and access rights | All projects collecting data on private land not owned by the data collector, or on public land | To respect the legal rights of landowners | Permission from private landowners; compliance with agency-specific land ownership and access policies for public land access and use |

Note: This is not a comprehensive checklist of all laws that may apply to your project, but rather a selection of laws that multiple citizen science projects have identified as relevant to their activities. Projects in each region or country are subject to different legal systems and should familiarize themselves with the laws of their own countries that address the issues targeted by these U.S. laws and policies. Additional considerations apply to projects operating across international boundaries.

a. Some federal projects have found ways around PRA compliance (see, e.g., Gellman 2015).

FIGURE 5.2. Land-access rights are an important consideration for citizen science projects. Not all land-use policies are communicated as clearly as those on this sign at a USDA research facility on public land. Nearby signs similarly specify land-use access conditions for pedestrians and bicyclists, with additional warnings of potential exposure to pesticides. Credit: photo by Andrea Wiggins.

Other legal and policy considerations are determined by data-collection protocols, such as the locations where data are collected. Land ownership and access rights play a significant role in where and when projects are legally able to collect data (McElfish et al. 2016). There are two types of land designation in the United States: government-owned land (or public land) and private land. Because public land requires permission for access by every type of entity (including individual, organizational, and governmental entities), while access to private land is at the sole discretion of the landowner, we focus on the nuances of public land access in this chapter. Projects that will collect data on private land must negotiate permissions directly with landowners or engage landowners as project participants.

Land owned by the government is considered to be held in trust for the public, providing recreational and natural resource value for the benefit of everyone, and is managed by myriad federal, state, and municipal authorities that designate the type of access and use of the land according to their agency missions (figure 5.2). For simplicity, we discuss federal agencies here. The mission of the U.S. National Park Service, for example, is to preserve the environment in perpetuity, which places restrictions on any type of resource extraction. On the other hand, the mission of the U.S. Forest Service is to provide sustainable yield to maximize the natural resource value of the land, which requires managed resource extraction. In the case of scientific research, these two agencies' differing missions can influence permissions and how research is done on public land.

If a study will collect specimens, the researchers will need to consult the policy for specimen collection outlined by the relevant land management agency before collection begins. For example, hobby paleontologists who are interested in collecting or cataloguing cultural resources on land managed by the Bureau of Land Management (BLM) will need to familiarize themselves with the Paleontological Resources Preservation Act. This act authorizes BLM to issue survey permits for research conducted on its land if the disturbance remains within one square meter of rock or soil. Furthermore, the permit also allows BLM to catalogue the research initiatives conducted on BLM land, which is vital information for long-term land-use planning.

When it comes to opportunistic or noninvasive data-collection techniques like recording observations via paper and electronic data-collection forms, identifying a policy becomes less straightforward, and projects should proceed with due diligence. For example, in 2015 the Wyoming State Legislature passed a law colloquially known as the Data Trespass Law, which was challenged in 2016 as unconstitutional (Bleizeffer 2016). The law makes noninvasive data-collection techniques illegal on private and public land within Wyoming and allows criminal trespass charges to be brought against individuals who engage in such activity without permission. This ongoing legal battle illustrates the complexities of data collection where public land and private land are adjacent and emphasizes the importance of determining land ownership before data collection begins.

The above information is helpful for projects seeking to comply with different laws and policies. Some projects may also wish to change or inform different policies, for example by collecting information to influence the formation of conservation policy or land management decisions. There are a number of laws and policies at the local, state, and federal levels that support public impact in decision making (McElfish et al. 2016). For example, through the Clean Water Act, the U.S. Environmental Protection Agency (EPA) allows citizen science data to influence a continuing planning process of priority setting and also to be used in evaluation. Similarly, the Endangered Species Act allows any person to submit a petition (to be accompanied by "substantial information") suggesting that a species should be listed as threatened or endangered. Species-specific laws, such as the Migratory Bird Act, are additional opportunities for input. Citizen science is increasingly promoted as a cost-effective mechanism for collecting the information required by these regulations (Evans et al. 2016).

While these laws do provide compelling opportunities for citizen science to be used in policy decisions, many also have important constraints. For example, if data are to be used by the EPA in decision making, they must be collected in line with agency-specific protocols. Such protocols are commonly made available on agency websites.

## Recommendations

After reviewing key laws and policies (table 5.1), projects are advised to:

survey international, national, state, and local laws that pertain to unique aspects of
their research, including laws and policies connected to funding source and to
planned data collection;

review their own organization's policies and those of any granting organizations,
especially to verify whether an ethics review is required from an IRB and/or
IACUC;

identify a few citizen science projects that have studied a similar topic and review
the specific methods utilized, the source of funding (e.g., BioBlitzes or long-
term ecological monitoring funded by the U.S. Geological Survey), and the data
policies of their websites to identify any additional considerations; and

become familiar—if the goal of the project is to advocate a change in land management policy or influence a public decision—with the applicable laws and policies
of the relevant authority (e.g., the laws of a federal regulatory agency or laws
pertaining to a specific species) (McElfish et al. 2016).

## Ethics and Human Subjects

In addition to the legal guidelines described above, ethical guidelines for the protection of
research subjects should also be considered. In citizen science, the role of volunteers is often
difficult to describe. In some ways, volunteers act as field technicians or as scientists who
participate in data collection, analysis, interpretation, and, in some cases, publication. In
other ways, volunteers act as research subjects. The observations they submit may contain
data about an organism and *also* contain data about the volunteer (e.g., by sharing a volunteer's location in real time). Additionally, projects may survey volunteers about their practices and preferences to support general research and individual project design. In both of
these cases, interactions with volunteers could be considered human subjects research.

Understanding what constitutes human subjects research requires understanding what
is considered research and who is considered a human subject. In the United States, the federal Department of Health and Human Services (HHS) sets the standards for human subjects research (see National Commission for the Protection of Human Subjects of Biomedical and Behavioral Research 1978). The HHS defines research as "a systematic investigation,
including research development, testing and evaluation designed to develop or contribute to
generalizable knowledge." Investigations designed to support operations are excluded from
this definition. In other words, a project manager who gathers survey feedback on tweaks to
a single aspect of project design—such as asking volunteers to perform activities alone versus in pairs—is not necessarily conducting research. But if that project manager later publishes an article describing the impact of socialization on data quality, the work is redefined
as research. Many (if not most) surveys of project participants that are intended to support

project operations also have potential to yield research results and benefit the broader community of citizen science practitioners.

Human subjects are defined as living individuals about whom an investigator obtains "data through intervention or interaction . . . or identifiable private information." For ecological citizen science, the most relevant portion of this clause is *interaction*, which designates any form of communication or contract between a researcher and a volunteer. This communication includes in-person as well as online interactions (e.g., asking a volunteer to take a digital survey about their project experiences).

In the case of projects conducted in a university setting, information about procedures for human subjects ethics review is typically available on university websites to help researchers determine whether an activity is considered subject to IRB approval. Because the definition of human subjects research is so ambiguous, project leaders who are unsure whether their activities fall under this umbrella are encouraged to consult these resources for clarification. Many IRBs may briefly review proposed research before officially designating this research as *exempt* from full review. Such exemptions are particularly common in social science research (e.g., if a project wishes to conduct anonymous interviews in order to better understand the motivations of citizen science volunteers). With that said, each IRB has independent decision authority, and what is considered exempt by one may not be considered exempt by another.

From an ethical perspective, all projects conducting human subjects research, whether funded or not, should align with ethical guidelines. From a legal perspective, all projects receiving federal funding (e.g., grants) are *required* to comply, as are many projects run by federal agencies. Citizen science projects operated by some, but not all, federal agencies must comply with these regulations as well.[2] Projects run by nonprofits should consult with their specific institutions.

Generally, human subjects research compliance requires three main categories of actions:

> submitting the proposed research for IRB review,
>
> ensuring that volunteers have the opportunity to give informed consent before participating in research activities, and
>
> providing a publicly accessible statement of compliance to relevant regulations.

Ethical guidelines for conducting human subjects research are provided by a range of international organizations, such as the World Health Organization, and by national authorities such as HHS in the United States. In some cases, there are additional guidelines in specific domains like medicine.

While there are no formal or broadly accepted ethical guidelines for citizen science researchers and practitioners at the time of this writing, established citizen science associa-

2. For a list of agencies that must comply, visit www.hhs.gov/ohrp/humansubjects /commonrule/.

tions are beginning to recognize the need for such guidelines. The European Citizen Science Association published a statement on "Ten Principles of Citizen Science" that touches on ethical concerns. The U.S.-based Citizen Science Association has an Ethics Working Group to develop insight and guidance on ethics in citizen science.

In the absence of specific guidelines, we recommend the Ethical Principles for Participatory Research Design (Bowser and Wiggins 2016), which were derived from the intersections of the Belmont Report[3] and guidelines for protecting privacy in participatory sensing. This framework was developed specifically for citizen science and includes seven principles that acknowledge the dynamic and diverse nature of citizen science. While several principles recommend adopting practices new to the field, others advocate following best practices that are already widely accepted and used by practitioners. We summarize these principles as follows:

> *Ethical engagement:* A project should identify a set of ethical principles (e.g., from established guidelines, professional codes of conduct, and project-specific goals) to govern project design, implementation, and management decisions.

> *Ongoing assessment:* The ethical principles must be available to multiple stakeholders so that each party can continually evaluate and improve practices supporting ethical engagement. Project leaders can demonstrate accountability and transparency by making their ethical principles explicit and available to funders, volunteers, and collaborators, which also improves the likelihood of adherence over time.

> *Informed participation:* Projects should emulate a process of informed consent to provide volunteers a clear and complete explanation of participation expectations prior to accepting contributions. Informed participation does not require increasing barriers to entry, but rather making project policies explicit and easily accessible throughout the participation process.

> *Evolving consent:* When project policies or participation expectations change, volunteers should be proactively informed, providing an opportunity to reconsider informed-participation choices. In other words, informed participation is not a one-time opportunity; evolving consent demonstrates respect for

---

3. The Belmont Report was issued in 1978 (and published in the Federal Register in 1979) by researchers working for the Department of Health and Human Services (HHS). The report was commissioned by HHS following controversy over the Tuskegee Syphilis Study, in which a disadvantaged population of African American sharecroppers were deliberately infected with syphilis and left untreated for the purposes of the research. The Belmont Report is considered an ethical standard within the United States and informs codes published by HHS and other federal agencies, including the National Science Foundation and the Office of Human Research Protections.

volunteers by acknowledging that changes in the project may impact their choice to contribute.

*Participant benefit:* Projects should find ways to maximize the benefits of participation to individuals and groups, which often means providing access to data and opportunities to engage more deeply in the science. While clearly intended to foster a mutually beneficial arrangement, participation is also a self-serving best practice: projects designed to explicitly deliver the experience, data, or outcomes that interest their volunteers tend to perform well overall. Providing easy access to project data is a common strategy to offer participants multiple ways to benefit from their involvement.

*Meaningful choice:* When informing volunteers about policies and expectations or changes to them, projects should clearly identify known or likely risks and benefits associated with participation. Such communication mirrors the standard informed-consent process in human subjects research but focuses primarily on supporting deliberate participation choices. Communicating the risks and benefits of participation in addition to the expectations and policies allows these choices to be more meaningful.

*Evolving choice:* Because contexts of participation may change in ways that affect volunteers' willingness to participate, projects should support ways to make contextually meaningful choices about participation. Providing this choice is intended to safeguard volunteers by giving them options for risk reduction related to disclosure—for example, enabling participation on different terms (or offering choices related to public access or visibility) for data collected at a public park versus a private residence. A common approach adopted by a number of projects is providing mechanisms for volunteers to obscure selected data points to preserve their privacy.

## Recommendations

Projects are encouraged to:

determine whether formal IRB assessment is necessary, based on the potential for producing *general* knowledge (such as that intended for peer-reviewed publications);

determine whether IACUC assessment is necessary, based on the level of involvement with nonhuman vertebrates; and

consider adopting the Ethical Principles for Participatory Research Design, which will include developing a short list of ethical principles to be shared with interested parties and identifying ways to ensure that volunteers have access to the policies, updates, and options that demonstrate respect for them as partners in the research or aligning with another ethics framework.

## RESEARCH LIFE CYCLE, PHASE 2: LAUNCH

During the second phase of the research life cycle, projects recruit volunteers and mobilize communities. In addition, protocols for data collection and quality assurance are piloted and refined, sometimes in conjunction with volunteers. From a legal, ethics, and policy perspective, the primary considerations during project launch are (1) the liability concerns that arise when engaging volunteers in data collection and (2) policies related to the use of technologies for collecting and sharing data. User agreements between projects and volunteers codify these considerations, along with expectations for data use.

### Liability and Risk Management

In U.S. law, *liability* is a legal obligation to pay debts or damages. In citizen science, *tort liability*—which occurs when the actions of a party cause damages to another party, who then files a complaint—is a key concern of projects and volunteers (Smith 2014). If a volunteer becomes injured in the process of collecting data, the project leaders or the organization hosting the project could face a lawsuit. Generally, projects manage this risk by having volunteers sign a liability waiver. While most projects involve minimal risk, this additional requirement may be important for some projects. Your legal counsel can help determine whether a liability waiver is necessary.

Researchers who study liability in sports and recreation (a similar context of voluntary participation) identify four factors that increase the likelihood that a liability waiver will hold up in court (Cotten and Cotton 1997):

Waivers should contain clear and unambiguous language.

Waivers should not codify expectations that breach law or policy, for example by asking volunteers to collect data on private land owned by other individuals.

Waivers must be presented in the form of a contract that both parties explicitly agree to. In other words, waivers hosted on static web pages are unlikely to hold up to scrutiny, which is problematic for many citizen science projects.

Waivers must be signed by parties who are legally capable of signing contracts (i.e., healthy adults over the age of eighteen).

These general guidelines for writing waivers would theoretically protect projects run in collaboration with different authorities, such as a nonprofits, government agencies, and/or academic institutions. Some researchers suggest that projects run out of federal agencies are additionally protected from liability through precedent, such as through court decisions related to the Federal Tort Claims Act. Projects that are self-organized and run by volunteers have a slightly different list of considerations and may find guidance in the white paper "Responding to Liability: Evaluating and Reducing Tort Liability for Digital Volunteers" (Robson 2012).

Policies and liability also need to be considered in regard to data collection. All projects need valid, high-quality data to achieve their goals. Data quality depends on data-collection protocols and on the tools and technologies used in data collection. In regard to protocols, the need to collect observational data with detailed metadata must be weighed against the potential risks of acquiring multiple and diverse types of information. When deciding which types of data and metadata to collect, we encourage project coordinators to ask the following questions:

1. What information is necessary to . . .

    a. Answer research questions or otherwise achieve project goals?

    b. Address similar research questions that support related goals?

2. Are there risks associated with the types of information collected?

    a. Is the process of data collection sensitive? For instance, might collection disturb vulnerable habitats? Likewise, are data collected on private land?

    b. Are the data themselves sensitive? For example, do the data contain information on the location of sensitive or endangered species?

    c. Are there additional data-collection concerns? For example, are there concerns related to land management rules?

3. What tools and technologies will be used to collect, upload, and share data?

    a. What are the risks associated with these tools? For example, many smartphones can automatically collect potentially sensitive location information.

Answering these questions often requires balancing different goals. For instance, a project may be required to choose between limiting data collection to protect volunteer privacy and collecting extra data to support future research. Recommendations for negotiating different goals are presented below in the section on managing, preserving, and sharing data. Regardless of which protocols are selected, coordinators have an ethical obligation to share these decisions with volunteers by making data-collection protocols clear and transparent, and also by implementing a user agreement.

## User Agreements

A *user agreement* is a contract between one organization (here, a citizen science project) and an employee, volunteer, or second organization. There are two main types of user agreements. *Clickwrap* agreements are explicit agreements between a volunteer and a project—either paper forms signed by volunteers or online forms where volunteers click a box saying "I agree" to a set of terms and conditions. By contrast, *browsewrap* agreements are implicit agreements. These take the form of posted policies that volunteers may see but do not explicitly agree to abide to.

Legal review of these two agreement types suggests that clickwrap agreements are significantly more likely to hold up in a court of law than browsewrap agreements (Organ and Corcoran 2008). Here, one determining factor may be whether volunteers are given *ade-*

*quate notice* and *meaningful opportunity* to review terms and conditions. One court ruled that the act of clicking "Continue" to move past an online form stating "I have read and agreed to [provider] terms and conditions" constituted a valid user agreement when the same form linked to those terms and conditions.

### Recommendations

Write a liability waiver stating that participation is voluntary, describing the potential risks and benefits of participation, and asserting that any damages incurred by volunteers are the full responsibility of those volunteers; ask volunteers to sign this waiver before participating.

Consider exactly which types of data you wish to collect and share, and communicate this decision to volunteers through protocols and terms of use.

Construct a clickwrap user agreement linking to information about participation, including a liability waiver, a description of data-collection activities, a data-use policy, and a privacy policy (some of these components are described elsewhere in this chapter).

If possible, seek legal counsel while crafting these policies, preferably during the pilot stage of project development, so that policies reflect any substantial changes to protocols or tools.

### RESEARCH LIFE CYCLE, PHASES 3 AND 4: MANAGE; PRESERVE AND SHARE

During the management stage of the research life cycle, projects maintain active data-collection and analysis activities. Data collection involves operational data-management activities—for example, to support a species observation record being transmitted from a volunteer's cell phone and ingested by a project's database. Preservation requires attention to data security and development of documentation to make data ready for sharing with external researchers and volunteers.

### Data Documentation

*Project metadata* is one form of documentation that describes key features of a citizen science project and associated data collection, including quality assurance protocols, the geographic scope of participation, units (e.g., meters or hectares), and descriptions of the data structure. This documentation, provided so that other researchers and volunteers can understand the types of data collected, rarely poses a risk to volunteers.

At the datum (or data record) level, *observational metadata* about individual measurements or observations are important for analysis and data sharing. Observational data may include information about a data point that is also indirectly about the volunteer who contributed the record, for example by stating the exact date, time, and location of data collection. Record-level data documentation has the potential to reveal details about volunteers' identities and must be handled with care (Bowser and Wiggins 2016). For NSF grantees, protecting volunteer privacy is a policy consideration: according to the NSF's grant policy manual, "privileged or confidential information should be released only in a form that protects the privacy of individuals and subjects involved" (National Science Foundation 2019). Protecting volunteer privacy is also an ethical best practice, as discussed above.

There are no universal solutions to the potential conflict between providing maximal privacy protection for volunteers and the growth of policies and practices supporting scientific ideals for open access to data that may include details on effort, location, and other information that could put volunteer privacy at risk. In cases where privacy threat is a realistic scenario, it is important to make a considered choice and document it in project policies. Ideally, decisions should be made through an ongoing assessment of values and concerns, conducted in cooperation with volunteers and other stakeholders (Bowser and Wiggins 2016).

In terms of documentation, the privacy policies presented on most citizen science project websites primarily cover the website itself (e.g., by describing whether cookies are used) and do not always mention how volunteers' privacy will be safeguarded when data are used for scientific research and related conservation, recreation, and education goals. Both types of information should be included, in addition to information on how data are de-identified or anonymized when shared. The "safe harbor" method of the Health Insurance Portability and Accountability Act,[4] developed to guide medical data sharing, provides a valuable set of guidelines for de-identification. The safe harbor guidelines emphasize the removal of eighteen unique identifiers and have been widely embraced within and outside of the medical domain.

Identifiers relevant to ecological citizen science may include the following:

Legal name

Social security number

All geographic subdivisions smaller than state

All elements (except year) of dates related to an individual (e.g., birthdate)

Telephone numbers, fax numbers, and email addresses

IP addresses

Device identifiers and serial numbers for mobile devices

Certificate/license numbers

Full-face photographs and comparable images

---

4. Not to be confused with the Safe Harbor Framework for complying with the privacy standards of the European Union (EU), which was developed in collaboration between the U.S. Department of Commerce and the EU.

TABLE 5.2 Levels of Privacy Risk for Different Types of Data

| Data type | Description | Example | Level of risk |
|---|---|---|---|
| 1. Raw data | Clearly identifiable data | Volunteer name, physical mailing address, email address | High |
| 2. Masked data | Use of pseudonyms | Pseudonym (or assigned identification number) replaces name but no other data are altered | High/medium |
| 3. Exposed data | Masking key attributes | Reducing GIS location precision to 1 km radius in publicly shared data | High/medium |
| 4. Managed data | Obfuscating key attributes *and* objective evaluation of risk | Replacing birthdate with birth year, in accordance with HIPAA and U.S.-EU Safe Harbor guidelines | Low/medium |
| 5. Aggregated/ anonymized data | Data contain no potential identifiers | Paper checklists of species occurrence data taken in a public park and returned to a drop box | Low |

Note: These levels are based on a synthesis of published information (El Emam 2010, U.S. Department of Health and Human Services 2015) and are applied here to the domain of ecological citizen science.

To clarify, we do not suggest that a project should never collect personally identifiable information or other sensitive data types like those discussed above. Rather, it is important to be aware of the potential sensitivities associated with different types of data, especially when setting up protocols and selecting tools or repositories for data storage, or when deciding whether and how to share data with volunteers, other researchers, and the public at large. To this end, researchers studying privacy have identified how different types of data correspond to different levels of privacy risk (table 5.2). Ultimately, it is important that data policies include information on which types of data are collected, and how these data will be used and shared. Such information might be placed on a page labeled *terms of use* or on a separate *privacy policy* page.

### Data Preservation

Like documenting data, preserving data serves multiple interests. One important consideration is supporting the scientific ideal of repeatable research. A second concern is supporting the expectation of volunteers that you will safeguard the data they invested their time and energy to contribute. Data preservation focuses on avoiding data loss in the short term, and on ensuring access to data in the long term.

Short-term data preservation policies are more likely to take the form of standard operating procedures for data backup and secure storage. While there are no widely known instances of citizen science projects encountering data-security problems, there is also no special immunity to viruses or identity thieves. Whenever possible, online data submission should be conducted through secure web forms to protect participants' privacy.

Long-term data preservation involves additional policy considerations. For instance, will data be deposited in a repository, and if so, are there any limitations that should be placed on access? Wherever reasonable, making data open access is increasingly required of publicly funded research, and it is also consistent with the goals of many citizen science projects. If data updates are regularly deposited, how will retroactive changes to data (editing or deleting data points) be handled? In most cases, for both practical and scientific reasons, deleted accounts or data are not removed from data that have already been archived, but are simply excluded from data updates moving forward.

### Recommendations

Consider exactly which types of data you wish to collect and evaluate how these data could threaten the privacy or security of volunteers.

If possible, decide which types of sensitive data you will obscure, and how.

Document decisions and potential risks in data policies, including terms of use or a separate privacy policy.

Decide how best to preserve your data in the short and long terms to balance security needs with potential benefits of broader use.

Communicate information to your stakeholders through data policies, such as terms of use.

### RESEARCH LIFE CYCLE, PHASE 5: DISSEMINATE RESULTS

Citizen science projects collect data to achieve a number of research, education, and policy goals. During the dissemination phase of the research life cycle, projects share raw and aggregate data, as well as analysis and interpretation, with different parties and at different times. In some cases, data collected by citizen science projects may be a form of intellectual property subject to copyright (Scassa and Huang 2015). *Intellectual property* is a legal term that protects the rights to "creations of the mind," including patents, creative works, and trade secrets. *Copyright* is a legal term that designates ownership of creative works. Within intellectual property and copyright law, there are two important considerations: potential copyright of individual observations as *creative works*, and potential copyright of project databases as *creative compilations*.

Some examples of creative works are literary works, including all types of written expressions; pictorial and graphic works, including photographs and drawings; and audiovisual works, including video footage and sound recordings (Scassa and Huang 2015). Many ecological citizen science projects collect information that may be considered creative works. For example, volunteers who photograph butterflies, record and share audio recordings of bird songs, and contribute to written expressions in the form of open-text comments or software could all be sharing copyrighted material. In the United States, copyright is automatically established when a new creative work is produced. Copyright gives the creator of a work the exclusive right to use and distribution that work for a finite period (currently, the author's lifetime plus seventy years). Copyright introduces restrictions on use that are not in keeping with the intent of many projects, but usage rights can be granted through strategies described later in this section.

According to the U.S. Copyright Act, creative compilations are produced through "collection and assembling of preexisting materials or of data that are selected in such a way that the resulting work as a whole constitutes an original work of authorship." In the context of ecological citizen science, databases that contain the observations of many volunteers could be considered copyrighted as creative works. However, as the 2001 Supreme Court Case *Feist Publications, Inc. v. Rural Telephone Service Company, Inc.* clarifies, there are some limitations to the copyright of creative compilations. Specifically, this case reached the Supreme Court when a small, rural publisher of a telephone directory (Rural Telephone Service Company) refused to share their directory with a larger, regional publisher (Feist Publications). The Supreme Court ruled that the Rural Telephone Service Company could not claim copyright of their directory for two reasons. Neither their method of data *selection*—namely, including all residents in a particular geography—nor their method of data presentation—intuitive alphabetical listing—involved sufficient originality to merit copyright protection. *Feist* also clarified that the factual data within a database, such as the street address of a particular individual, never enjoys copyright protection.

To date, no legal cases have addressed whether citizen science datasets should be considered creative compilations, but legal analysis suggests that some citizen science datasets may be considered creative compilations and thus subject to copyright law (Scassa and Huang 2015). In such situations, there may be an inherent conflict between the copyrights of volunteers, for example when original photographs are submitted, and the copyrights of researchers who create compilation databases. In such a scenario, it is unclear whose rights might take precedence.

Any potential conflict may be resolved through data policies and browsewrap user agreements that make these policies binding. Citizen science volunteers may formally waive their copyright by transferring copyright to a project or by placing an individual creative work in the public domain, or they can apply a license that grants permission for use. Licensing may be used when volunteers or projects wish to maintain copyright while sharing data or other information. A *license* grants formal permission for one party to access the intellectual property of another and dictates the acceptable terms of use. For example, a project might permit reuse of its data for research and education, but not for commercial use. Or a project might require that a copyright notice linking to the project's web page accompanies all instances of reuse. Creative Commons produces a popular collection of international licenses designed to support data sharing under various conditions, available from its website (https://creativecommons.org). The Creative Commons website also hosts an online tool to generate customized licenses that typically include the type of work (e.g., a dataset), the name (e.g., of a project), and a URL.

In some cases, a project may wish to completely waive ownership of its data. Designating a dataset part of the public domain, which means forfeiting all benefit from and control over the product, may achieve this end. However, projects that put their data in the public domain may still be under certain obligations, such as the moral right to object to abuse of the data, and Creative Commons recommends against placing copyrighted data in the public domain for precisely this reason.

Many project data policies should therefore assert ownership of the datasets collected and also describe conditions for reuse. This information supports volunteer agency in deciding whether to contribute to a project and helps other researchers discover and use datasets. Furthermore, it is equally important for data users and projects that integrate third-party datasets to consider the stated policies of data providers, where available. For example, while most U.S. government data (such as census data) are designated as open for all to access and reuse, data collected through private-sector institutions or specific government entities (e.g., the Ordnance Survey in Great Britain) may place restrictions on reuse. Some citizen science projects simply require properly citing datasets. Other projects ask that a project lead be included as a coauthor on any publication resulting from data use. These policies should be adhered to as best practices from an ethical point of view, and they may also be legally binding.

We do not mean to suggest that all projects must share data through the mechanisms described above. We recognize that the level to which each project can and should share data, both with other scientists and with the general public, is highly variable and depends on many contextual factors. On one hand, data sharing promotes new knowledge and is increasingly considered a best practice or even required. For example, the NSF states: "Investigators are expected to share with other researchers, at no more than incremental cost and within a reasonable time, the primary data, samples, physical collections and other supporting materials created or gathered in the course of work under NSF grants" (National Science Foundation 2019). On the other hand, many researchers rely on publication to support their careers. In such cases, the success of current and future citizen science projects may depend on keeping raw data private, at least until the results of a study can be published. Researchers should also resist sharing unprocessed datasets when doing so could put participant privacy at risk—a point acknowledged by the NSF in their data policy manual. Producing a modified dataset for distribution purposes is often a reasonable compromise when these concerns are in tension.

## Recommendations

Projects that may be collecting copyrighted information from volunteers should use data policies to indicate whether volunteers are waiving copyright or applying a license to their data through participation.

Projects that believe their datasets are creative compilations should decide how, when, and whether volunteers, external researchers, and the general public should access data; decide how data should be cited or attributed, and identify a suitable license; and communicate this attribution and license information to volunteers and potential data users through data policies, such as terms of use, and in descriptive documentation of the dataset.

Data users and projects that wish to integrate their data with external sources should carefully review the associated data policies, including terms of use, of partner researchers and their host organizations.

### RESEARCH LIFE CYCLE, PHASE 6: MEASURE

During the measurement phase of the research life cycle, projects assess the effects of citizen science activities and document important outcomes. Periodically reflecting on the role of various legal, ethical, and policy considerations can help researchers improve the design of future citizen science activities and related policies. In addition, sharing assessment and evaluation information with other practitioners, whether through publication or communities of practice, can help support and advance the practice of ecological citizen science. As previously mentioned, however, assessment data may also be sensitive, particularly if they includes such material as volunteer-feedback survey responses. One important community of practice, the Citizen Science Association (www.citizenscience.org), hosts a biannual conference to convene practitioners and support information sharing and collaboration. Sister organizations in Europe and Australia provide similar opportunities for international knowledge sharing.

### SUMMARY

This chapter presents important legal, ethical, and policy concerns associated with ecological citizen science. During project planning, it is important for coordinators to understand all applicable regulations connected to institutional affiliations, sources of funding, and data-collection plans. Project coordinators hoping to influence policy decisions should also align their protocols with existing laws and standards to maximize the usefulness of their data. Launching a project often requires mobilizing a significant number of volunteers and dealing with potential liability concerns through the use of waivers. The practical decisions involved in managing a project, particularly around data preservation and sharing, require balancing ideals of openness with volunteer privacy protections. Disseminating results and sharing data invokes intellectual property concerns, as both individual contributions and databases as compilations may be copyrighted under U.S. law. Sharing evaluation outcomes can both benefit the practitioner community and require another set of decisions around data sharing.

None of the aforementioned challenges are insurmountable. However, projects are encouraged to take the information in this chapter as a starting point and educate themselves on the key legal, policy, and ethical considerations that come into play at each stage of the research life cycle. Planning helps maximize the value of the data that a project collects and analyzes, and it should be considered an iterative process that includes and takes advantage of feedback from project volunteers. In addition, clearly written policies should communicate information ranging from the privacy implications of data-collection protocols to the intellectual property rights of projects and volunteers.

# SIX

# Recruitment and Building the Team

MICHELLE D. PRYSBY

Finding and engaging participants in your citizen science project is an important part of the process. After all, if citizen involvement did not matter, you would be pursuing your research, conservation, or education goals in a different way. There is not, however, any one formula for volunteer recruitment. Your project could involve ten volunteers or ten thousand, young children or expert naturalists, all in one neighborhood or scattered across the continent. To illustrate the breadth of possibilities, let us examine three very different citizen science projects that are all considered successful.

In 1999, two scientists, Doug Wachob of the Teton Science School and Bruce Smith of the National Elk Refuge, launched a study of elk migration patterns near Jackson Hole, Wyoming. In 2001, while presenting some preliminary results to the community, they realized that residents had local knowledge and observations to share about elk movement through their neighborhoods and that they might be able to learn more by involving the residents in collecting additional observations. The two scientists thus revised their study to include a citizen science component. Wachob and Smith then selectively chose homes to maximize the geographic area covered and contacted homeowners individually to invite them to participate. The scientists' team provided in-person, individualized training at each resident's home and also made personal visits to all twenty participants to retrieve the data at the end of the study. While the citizen science volunteers did not detect any new significant elk migration routes, they did provide additional details on nonmigratory movements of elk and validated the data collected by the field technicians (Wachob and Smith 2007).

A second example comes from Snapshot Serengeti, an online citizen science project that is part of the Zooniverse suite of projects. Snapshot Serengeti volunteers identify wildlife through images taken by infrared cameras in the Serengeti National Park. The scientific goals of the project are focused on understanding how competing wildlife species coexist in a landscape. The assistance of volunteers is extremely useful in order to process the vast amount of data produced by the more than two hundred cameras deployed on the landscape. A typical field season results in tens of thousands of photos to be viewed and interpreted, and each photo is interpreted by multiple volunteers in order to maximize data quality. Unlike the Teton elk project, Snapshot Serengeti requires a vast number of volunteers, but they can be located anywhere in the world with an Internet connection. Hundreds of thousands of people have participated in the project online (Swanson 2014, University of Minnesota Lion Project 2014).

The third example illustrates yet a different type of volunteer citizen scientist. Thousands of children have helped scientists track the migrations of monarch butterflies, hummingbirds, bald eagles, and other species each year since 1994 through the Journey North program. The children contribute by reporting the first sightings of these species in their area. Although any member of the public can report observations to Journey North, the program is especially tailored to schoolchildren, with classroom lessons designed to teach students about phenology, global change, ecology, conservation, and more while engaging them in collecting and sharing real data. Like Snapshot Serengeti, Journey North involves very high numbers of volunteers, but it is not solely an online project, as the volunteers are contributing their own observations of monarchs and other species in the field. Journey North has a strong focus on providing programs and resources to improve K–12 education. Thus, developing curricula that help teachers incorporate citizen science activities into their teaching is a cornerstone of the project and an aspect that helps attract their key audience of K–12 students (https://journeynorth.org).

Each of these three programs used quite different strategies to recruit volunteers. The elk project required going door-to-door and issuing personal invitations to residents, because the study area was small and the project had goals related to engaging the local community. Snapshot Serengeti, on the other hand, focused on recruiting volunteers through social media and

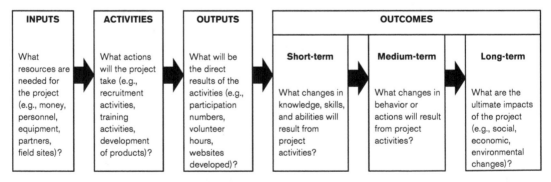

FIGURE 6.1. An example of a logic model used to define project goals and outcomes for a citizen science project. To interpret the logic model, read it from left to right, answering the questions with a series of "if . . . then" statements. For example, "If we have and use these resources, then we can conduct these activities. If we conduct these activities, we will produce the following outputs. With these outputs, we will achieve these short-term outcomes." Credit: based on Kellogg Foundation (2004).

other online methods because of its need for many volunteers. Finally, Journey North focuses mainly on word-of-mouth and online recruiting, but it originally connected with K–12 students by partnering with their teachers and school districts. Despite these differences, each project has achieved its intended outcomes related to both science and public outreach.

Because the degree to which a citizen science project can be accomplished depends on its volunteers, a central part of any project is recruiting participants. Thus, the focus of this chapter is the recruitment process. Notably, even if you are part of an existing project and already have the volunteers you need, you still will want to consider other aspects of building the team, such as articulating your project's goals, understanding the motivations of your volunteers, and including a variety of expertise on your project team, all of which are discussed in this chapter.

## UNDERSTANDING YOUR PROJECT'S GOALS AND NEEDS

Before crafting your volunteer recruitment plan, it is important to understand and *put into words* your project's goals and intended outcomes (chapter 4; figure 6.1). After you have developed and articulated your project's goals, you can develop a better understanding of the volunteer audience you need to engage for recruitment. Some questions to consider include the following:

How many volunteers do you need to involve to achieve your research, education, and conservation goals?

Do you need volunteers with a particular skill or expertise? If so, is it a skill that can be developed through training you offer, or do you need to seek volunteers who are already skilled?

Do you need the volunteers to work in a specific geographic area?

Will the volunteers need to identify and have access to their own field sites?

Do you need volunteers who will work independently or in groups?

Do you need volunteers who will participate in the project repeatedly over time, or can participation be one time only?

Do your educational goals indicate a particular audience, such as K–12 students or teachers, landowners, or underrepresented communities?

Will the volunteers be engaged solely in data collection and reporting, or will they be involved in planning, developing protocols, and other aspects of the scientific process?

What level of autonomy will volunteers have in the project?

Do the volunteers need access to and the ability to use technology, such as the Internet?

What are the minimum and maximum expected time commitments for a volunteer?

Once you have defined the volunteer audience you wish to engage, consider how best to connect with them. The following questions can help guide you:

What media sources does this audience prefer?

To what organizations do individuals in this audience belong?

What motivations would this audience have to volunteer?

Does this audience already have a relationship with you or your organization? Is it a positive one?

Who are the trusted leaders or respected members of this audience?

What barriers might exist for this audience in terms of volunteering with the project?

What incentives might appeal to this audience?

## UNDERSTANDING VOLUNTEER MOTIVATIONS

In looking at the diverse array of citizen science projects now in existence, it is truly astounding what volunteers are interested in and willing to do. For instance, they collect water samples and plate them in their homes to look for fecal coliform bacterial growth (Stepenuck et al. 2010), commit to checking and reporting on weather data at their homes every single day for years (National Oceanic and Atmospheric Administration 2019), and spend hours on the computer, painstakingly transcribing handwritten museum labels into an electronic database (Allen and Guralnick 2014). Others subject themselves to biting insects and thorny plants to reach field sites, rise before dawn or stay out late at night to collect data, and endure all kinds of inclement weather. In other words, on the surface it appears that volunteers will do almost anything in the name of science and conservation. However, gaining a better understanding of volunteer motivations can help a citizen science project build a cadre of volunteers who are best suited to the project's goals and who will actively participate and continue with the project over time. In fact, research demonstrates that matching volunteers' needs and motivations to the project's goals will improve volunteer recruitment, participation, and retention (e.g., Clary et al. 1992, Houle et al. 2005). Hence, matching volunteers' needs and motivations can save time, money, and other resources for the project and increase the likelihood of achieving its goals.

One of the primary reasons people volunteer for anything is that they were asked (Freeman 1997). However, while simply asking may get volunteers in the door (or out in the field), citizen science volunteers have their own motivations for participating. One primary motivation for participation in ecological and conservation-related projects is a desire to help the environment. There are also social motivations; a desire to learn new things, to spend time in and experience nature, or to alleviate guilt about environmental issues; and potential career benefits (Grese et al. 2001, Bruyere and Rappe 2007, Jacobson et al. 2012). These motivations may change over time, and motivations that are important initially may not be the factors that predict long-term commitment to a project. However, social motivations— the desire to meet and spend time with people who have similar interests—are one of the best predictors of volunteer commitment, which includes the frequency, duration, and extent to which a volunteer continues to participate over time (Ryan et al. 2001, McDougle et al. 2011, Asah and Blahna 2012).

Much of our understanding of volunteer motivations for citizen science projects has focused on online, virtual projects (particularly projects in the Zooniverse suite) rather than field-based projects. As with other ecological and conservation-related volunteering, research shows a wide variety of reasons why volunteers participate, ranging from personal motivations, such as a desire to satisfy one's curiosity, to more altruistic motivations, such as a desire to contribute to a large scientific endeavor (Rotman et al. 2012, Raddick et al. 2013, Reed et al. 2013). Again, initial motivations for contributing to citizen science efforts may differ from the factors that predict commitment over time (Nov et al. 2011, Eveleigh et al. 2014, West and Pateman 2016). For example, one study found that participants' initial motivations for joining citizen science projects were focused on personal interest, but other motivations, such as a desire to help scientists, became more important for continued participation (Nov et al. 2011).

The bottom line for citizen science project organizers in ecology and conservation is the importance of taking time in the project development phase to identify and understand the potential motivations people may have to volunteer with the project, and then design the project to fill those needs and desires as much as possible within the context of the project's scientific, conservation, and educational goals. Consider designing the project to meet a wide range of motivations by having a variety of roles, levels of participation, and activities from which volunteers may choose. You can gather information on motivations and interests through focus groups, surveys, talking to key informants, or looking at similar projects and efforts. Again, keep in mind that volunteers' motivations may change over time. In particular, consider the importance of social aspects over time, and consider designing the project to promote social connections among volunteers.

## VOLUNTEER RECRUITMENT STRATEGIES

At the very beginning of the project, consider involving volunteers by having a small group of them as part of your project planning team, providing advice and perspectives as you develop goals, protocols, training plans, and so on. This inclusion at the planning stage is especially helpful if you are hoping to engage volunteers from marginalized communities or from groups traditionally underrepresented in the environmental field.

A good rule to practice is to recruit volunteers only when you have something ready for them to do. A large lag time between recruiting volunteers and actually having specific activities for them can result in volunteers losing interest and moving on to other priorities in their lives. You should be prepared to handle inquiries about volunteering year-round.

You may not need new volunteers until the summer field season, but perhaps a prospective volunteer runs across your website in December, is excited by your project, and sends an email asking how to volunteer. Do not let the inquiry languish, thinking that you will send a response in the spring. Be prepared to respond to all inquiries from prospective volunteers promptly, letting them know the timeline for the project and when they can expect to hear from you again. Volunteer recruitment is a year-round process. Even when you are not actively signing up volunteers, you will want to be doing other activities (e.g., website updates, newsletters, press releases, talks for community groups) that will raise public awareness about your project. Then, when you are ready to recruit volunteers, you will have a pool of people who are already familiar with and, hopefully, interested in the project. Even the most well-run citizen science projects face volunteer attrition, so recruitment must be done repeatedly if the project is anything beyond a one-time event.

For prospective volunteers to evaluate whether an opportunity fits their needs and interests, they need to know the specifics of what they are being asked to do. Develop a "job" description that states exactly what the expectations for volunteers are, including the purpose of the project, the expected time commitment, required skills, specific activities and where they will take place, and any training needed or provided. Highlight the expected benefits of participation, which ideally will match the motivations you expect your target volunteer group to have. Ultimately, the project's goals, the number and type of volunteers needed, the project's geographic scope, and the resources that are available will all determine the specific recruitment strategies you use.

### Door-to-Door Personal Contact and Neighborhood Recruitment

Going door-to-door to talk with individuals is time-consuming, but it may be the best method for projects that are focused on a small geographic area, that need to engage homeowners, or that aim to engage a marginalized population such as residents of a low-income neighborhood. The elk project described at the beginning of this chapter is a good example of a project that needed the buy-in of numerous landowners within a small and defined geographic area in order to be scientifically successful and that also had goals related to engaging the local community around the education center leading the project. A similar recruitment strategy can be found in another form of participatory research, popular epidemiology. Popular epidemiology projects often begin when community members recognize a common health issue in the community. The community members work through grassroots efforts to develop maps of where these diseases or health issues are occurring and locations of possible causes such as sources of pollution, and they go door-to-door to engage other members of the community in the efforts (Leung et al. 2004).

To build trust with community members, it can be effective to establish a relationship with a local community leader, such as a religious leader, tribal leader, or cultural practitioner. These leaders can help you better understand the needs of the community, the best methods for recruiting and communicating with community members, and how your project's goals could mesh with the community's own goals (Bodin et al. 2006). Working with these leaders is particularly important when you are not already part of the community you wish to engage, or when there is any history of distrust or animosity between the community and your institution.

Once a small group of individuals are on board with your project, you can consider a *snowball* recruitment technique, asking each individual to recruit additional friends, family, or neighbors to participate. Be aware, however, that snowball recruiting can lead to a homogenized volunteer base if the initial group of volunteers is not racially, ethnically, geographically, or socioeconomically diverse.

Another technique for recruiting volunteers from a defined geographic area (such as a neighborhood) is to work with existing organizations that either have their own members or have ties to the community you wish to reach. For example, neighborhood and homeowners' associations typically have established communication methods such as community meetings, newsletters, and listservs that could possibly publicize your project. In addition, the elected leaders of these organizations can assist you with identifying potential volunteers.

### Collaborating with Other Groups and Organizations

Collaborating with existing organizations is one of the most effective ways to involve volunteers in citizen science. These collaborations have the added benefit of providing infrastructure that can make communication, training, data sharing, and volunteer retention easier. For conservation projects in the United States, Master Naturalist (e.g., https://anrosp.wild-apricot.org; Larese-Casanova and Prysby 2018) and Master Gardener (https://mastergardener.extension.org) organizations can be very good resources. These are lifelong learning programs that usually have a required volunteer component for participants, so they are often seeking potential projects for their volunteers. The participants take an extensive (often forty or more hours) training course on natural history and natural resource management (in the case of Master Naturalists) or on home horticulture (in the case of Master Gardeners). The participants typically (though not always) have ties to the Cooperative Extension Service (the arm of land-grant universities in each state that is focused on conducting and sharing applied research for the betterment of individuals' lives and communities). Cooperative Extension offices typically have educators and volunteers working in many or all counties within a state, making them an excellent community resource.

A good example of a Master Naturalist program is Virginia's (www.virginiamaster naturalist.org), which is sponsored by Virginia Cooperative Extension (associated with Virginia Tech) and has program chapters in thirty localities throughout the state. Virginia's program has more than 1,800 active volunteers, all of whom engage in regular volunteer work in natural resource education, stewardship, and citizen science. In one example of a citizen science partnership, scientists at Virginia Commonwealth University partnered with the Virginia Master Naturalist program to engage volunteers in monitoring vernal pools in their communities. By working with the program leadership, the scientists connected with volunteers who were both interested in vernal pool monitoring and located near the identified vernal pools. Furthermore, the infrastructure of the Virginia Master Naturalist program, including locally based chapters with elected leaders, communication systems, and incentives for volunteering, helped maximize volunteer retention in the project, ensure that data were submitted, and provide additional project publicity.

Another type of organization that can be a wonderful partner and source of volunteers is a "Friends of" organization. Many parks, wildlife refuges, community centers, and libraries have such groups. For example, Friends of Sky Meadows State Park, in Virginia, conduct bluebird nest-box monitoring at the park. The participants submit the bluebird data not only to the park, but also to a statewide organization (the Virginia Bluebird Society) and even to a national citizen science project (Cornell Lab of Ornithology's NestWatch). Other interest groups that could be partners for volunteer recruitment include bird clubs, wildflower or native plant societies, and herpetology clubs. These groups are made up of both professional and amateur naturalists, some highly skilled.

Natural resource recreation groups such as hunting and fishing organizations, hiking clubs, and boating clubs are another set of community or interest groups to consider. For example, the Appalachian Trail MEGA-Transect project, a collaborative effort to use the hiking trail as an environmental monitoring transect, regularly engages volunteers from Appalachian Trail hiking clubs in collecting data for studies of American chestnut populations, phenology, and invasive species (Appalachian Trail Conservancy 2019). Likewise, state wildlife agencies engage hunters in measuring populations of American woodcock, ruffed grouse, and other game species of interest. Sometimes the group may be an unexpected one. For example, the Lion Guardians project in Kenya collaborates with Maasai men who traditionally kill lions to protect their livestock. Through the Lion Guardians project, the men are instead incentivized to monitor and protect lion populations (Hazzah et al. 2014).

Youth-based interest groups, such as scouting programs and 4-H, are often eager to have the youths participate in meaningful volunteer service and could be partners for citizen science projects that focus on youth participation. Partnering with after-school programs or clubs is one way to engage young people while avoiding some of the difficulties of designing a project to match the existing curriculum requirements and learning standards of a formal classroom.

## Collaborating with Professional Groups

In some cases, it may be more effective to solicit participation in your project from individuals in specific jobs or professions. These individuals could be participating in the project voluntarily, but also as part of their regular work. A long-running project that uses individuals from specific professions is the Rural Mail Carrier Survey. Since the 1940s, state wildlife agencies have asked rural mail carriers to report wildlife sightings as part of the agencies' tracking of game species' population trends (Robinson et al. 2000). Since the rural mail carriers are already driving roads in consistent patterns all across the states, they represent an inexpensive and far-reaching volunteer pool. More recently, the U.S. Geological Survey is engaging professional rafting guides in a citizen science project to monitor aquatic insect populations on the Colorado River in the Grand Canyon. As with the rural mail carriers, the rafting guides are already in the right place at the right time. The guides are eager to incorporate the monitoring into their trips, because it provides an interesting camp activity for their passengers, and the guides themselves care deeply about the Grand Canyon and are interested in helping with studies that increase understanding of this ecosystem (U.S. Geological Survey 2019).

### Collaborating with Non-formal Education Centers

Non-formal education centers, such as nature centers, science and natural history museums, environmental education centers, and botanical gardens, often make excellent partners for citizen science endeavors. They typically have members who could be engaged as volunteers, and they often run K–12 school or after-school programs into which citizen science projects could be incorporated. Moreover, these non-formal education centers have staff with expertise in education, and they often have land that can make suitable field sites, or else have easy access to public lands.

One program that has partnered effectively with non-formal education centers, particularly nature centers, is the Monarch Larva Monitoring Project (https://monarchlab.org /mlmp). One of the primary ways the project recruited its original volunteers was through trainings held at nature centers in key geographic areas. Since that initial recruitment, the project has expanded to partner with additional nature centers to engage youth and other audiences through outreach programs and training workshops.

The USA National Phenology Network and its Nature's Notebook program (see chapter 14) have used partnerships with non-formal education centers to support their educational goals and to develop local research studies that tie into their national-level research. In Tucson, Arizona, they collaborated with more than a dozen education centers, including parks, botanical gardens, and environmental education centers, to develop a Phenology Trail, essentially a set of sites with shared research questions and educational curricula. Each site tailors the project to fit its own education and outreach programs but maintains similar enough protocols to contribute to the larger project. This effort is still relatively new, but preliminary evaluations indicate that the collaboration is having a positive effect on volunteer recruitment and retention for the Nature's Notebook project, and they are expanding the Phenology Trail concept to other communities (USA National Phenology Network 2012, 2015).

### Collaborating with Classrooms

Involving K–12 students in citizen science is both rewarding and challenging. With increasing focus on standardized testing and accountability, it is often challenging for teachers to add extra things to the curriculum. To effectively engage K–12 classrooms in a citizen science project, your program should document how the project is tied to the specific knowledge and skills taught at a particular grade level and how involvement in the project will improve students' learning outcomes. Also keep in mind the end users of your data and whether they will find data collected by children credible. That said, many successful citizen science projects have focused on engaging K–12 students, such as Journey North (https:// journeynorth.org), Project Budburst (http://budburst.org), and the Lost Ladybug Project (www.lostladybug.org).

The most efficient way to connect with K–12 classrooms is through teachers. A common strategy is to offer professional development for teachers to train them in the curriculum associated with the project. Professional development workshops can be offered independently, through the school district, or at conferences attended by teachers, such as the National Science Teaching Association's conferences, or state-level environmental education conferences. Partnering with teacher education faculty within a school or department of education at a college or university when planning professional development for teachers also may be effective.

College and university students are another potential pool of volunteers. You can partner with faculty or leaders of campus student organizations to engage students in citizen science to enhance their studies and educational experiences. In one example of such a collaboration, the Beetle Project at SUNY Empire State College aims to increase student retention in science disciplines while also improving the understanding of Japanese beetles in New York. Multiple courses at the institution incorporate the project into the curriculum, using it as a real-world study in which students can practice all aspects of the scientific process (Empire State College 2016).

### Event-Based Recruiting

Some projects recruit volunteers through special events, such as festivals, conferences, or public outreach events. Staffing a booth at a community event, such as an Earth Day festival, is an opportunity to disseminate information about your project as well as to pique interest in becoming a citizen science volunteer. Use a public lecture or field trip to draw in people interested in lifelong learning, and then invite them to participate in the project. Partnering with a nature center or park is particularly useful for this strategy, as they likely can assist with hosting and advertising the event.

### Recruiting through Mass Media and Social Media

Depending on the geographic scope of your project, both traditional mass media and social media may be useful recruiting tools. If you would like to use traditional mass media, develop a media contact list for the geographic area of your project, to include key subscription-based newspapers, free newspapers, public and commercial radio stations, and public and commercial television stations. Send press releases about your project to your media contact list, submit public service announcements to the radio stations, and invite local TV news to your events. Depending on your project's budget, you could consider paid advertisement on websites and in publications read by the specific audience you wish to engage.

There are online listservs, discussion boards, and user groups for nearly every interest group imaginable. Circulating your recruitment materials through these groups is generally free and simple to do. Use of social media (e.g., Facebook, Twitter, LinkedIn, Instagram) to recruit volunteers will be most effective if you have already built a following of social media users by creating and sharing high-quality content. Starting a brand-new Facebook page for your project is unlikely to yield many new volunteers, but sharing information about your project on pages that already have a lot of followers can. Building a social media following takes time and dedication to posting new and captivating content regularly. Either begin the process of building a social media following well before you need to recruit volunteers or instead make use of the social media presence of your agency or organization and any partnering organization. Keep in mind that you can link different social media platforms together so that, for example, a new blog post results in an automatic Facebook profile update and an automatic tweet. You also can make use of paid advertisement options on social media sites such as Facebook. There is little research into the effectiveness of recruiting volunteers through paid social media advertising, but one benefit of this strategy is that you can often target your ads to very specific audiences (based on zip code, age, gender, etc.)

One crowdsourcing-type citizen science project that made good use of traditional and social media was Galaxy Zoo, the first citizen science project in what is now the Zooniverse suite of projects. In 2007, an initial post about the project on one of the project organizer's websites, combined with a press release picked up by BBC television, yielded tens of thousands of people registering as volunteers on the project website within just a few days (Clery 2011). While it was likely a combination of factors that helped the Galaxy Zoo project go viral, it certainly helped that one of the project leaders was a co-presenter on a long-running BBC television documentary show on astronomy and thus had an existing social media following and media contacts.

### Recruiting through Volunteer Referral Sites

Volunteer centers and volunteer referral sites serve as clearinghouses for volunteer opportunities at local, state, or national scales. VolunteerMatch.org, for example, is a national-level site where organizations can post volunteer opportunities, and potential volunteers can filter the opportunities based on location, topic, and other variables. Similarly, the United Way operates volunteer centers on a local community scale.

The most significant volunteer clearinghouse for citizen science is SciStarter (https://scistarter.org), an online portal where citizen science project leaders can post information about their projects and interested volunteers can seek out projects by topic, location, or type of activity. In addition, SciStarter promotes individual projects through blog posts and an e-newsletter.

The EarthWatch Institute (https://earthwatch.org) is well known for its strategy of recruiting citizen science volunteers through expeditions—travel experiences that involve participants in doing fieldwork for specific research studies. EarthWatch participants pay to participate and join the expeditions. In this way, EarthWatch is able to act as a grant-making entity, providing funding to scientists to lead the research as well as a volunteer recruiter for the projects (EarthWatch Institute 2014).

As with mass media recruitment techniques, volunteer referral sites cast a very wide net, and you should include a clear "job" description and possibly screen potential volunteers to be sure that they meet your project's needs. The level and type of screening necessary will depend on your own organization's policies and the volunteer roles. Your organization may require a high level of screening if the volunteers are considered unpaid staff, if they are in a position of trust, and if they are protected by the organization's liability insurance.

## CONSIDERING DIVERSITY AND INCLUSION

*Diversity* among citizen science participants can refer to many attributes, including race, ethnicity, cultural background, income level, socioeconomic status, geographic location, gender, educational background, level of scientific or naturalist expertise, and more. Like many facets of the natural resource profession, ecological and conservation-oriented citizen science projects often have volunteer bases that lack diversity in one or more of these areas (Porticella et al. 2013, Taylor 2014). This lack of diversity may not be a hindrance to accomplishing the scientific goals of a project. A pool of volunteers that all have strong naturalist skills, experience and comfort in the outdoors, college degrees, and a lot of leisure time may indeed be easy to train for the project and easily able to collect high-quality data and participate fully in other aspects of the scientific process. They may, on the other hand, be primarily white, English-speaking, and middle or upper middle class.

There are a variety of reasons why citizen science practitioners may want to diversify their volunteer pools and make their programs more inclusive. Projects based in or sponsored by federal or state agencies often have a mandate to reach diverse audiences and must report demographic data. If you are organizing a project with a narrow geographic scope, and a diverse population lives in that area, you will likely want to attract all sectors of that area. The project's educational or conservation goals may mean you want to involve volunteers with varying backgrounds in order to more significantly impact participants' learning and behaviors. A diversity of backgrounds could lead to new insights and outcomes that would not otherwise be possible.

Diversity is a topic of interest in both the citizen science and the natural resource, ecology, and conservation communities. While there are no formulas for increasing diversity that work across all projects, there are a few recommendations to consider. Keep in mind that some of the recruitment strategies described earlier in this chapter (snowball, working with preexisting bird or wildflower clubs) tend to lead to less racial, ethnic, and socioeconomic diversity. Methods that could be more effective at promoting diversity include working with schools that serve diverse neighborhoods, partnering with religious or other community leaders, and providing benefits to participation that are well aligned with the interests of the community (Robinson 2008, Porticella et al. 2013). A 2013 report commissioned by the Association of Science-Technology Centers highlights many specific strategies and examples of programs that have focused on engaging diverse audiences (Porticella et al. 2013; see box 6.1).

## REMEMBER OTHER TEAM MEMBERS

Depending on your own skills and experience, you may need to recruit individuals with specific skills who can assist with project vision, development, and execution. A robust project team includes members with expertise in a variety of scientific skills, including project design and planning, protocol development, data management, information technology, and statistics. Likewise, the project team should also include members with expertise in education and outreach skills, including volunteer management, developing and presenting training materials, developing curricula, communications, public relations, and evaluation. In addition, if you are a scientist or land manager designing a top-down-type project, it is a good idea to involve as part of the project team, from the beginning, one or more leaders from the community you hope to engage. These community leaders can help ensure that the project will be a good fit with the community and inform decisions about protocols, communication, and training to ensure that they are effective. Likewise, if you are part of a community group developing a grassroots or bottom-up citizen science project, consider including scientists, land managers, and other subject-matter experts as part of your planning team or as an advisory board, particularly if their areas of expertise are not otherwise represented in your community group. A scientist can assist you in developing research protocols and quality-control systems that will result in meaningful and statistically sound data.

## SUMMARY

The most important step to building a successful team is having a strong understanding of your project's goals and an ability to articulate them clearly, be they scientific, educational, or conservation-oriented. A thorough understanding of what you are trying to achieve will clarify the expertise you need on your team. In thinking about the volunteer component, you will want to understand the likely needs, interests, and motivations of your potential volunteers so that you can make sure your project will be a good fit. Your specific recruitment strategies will match your intended volunteer audience.

If improving science learning for youth is an important goal, you will want to pursue collaborations with and recruitment strategies for K–12 classrooms or non-formal youth groups. If one of your goals is changing landowner behavior in a specific area, you may need to knock on doors to recruit individuals within that area. If the success of your project depends on massive participation but a low time commitment per volunteer, you may instead be looking at social media or other online recruiting tools.

In the end, keep in mind that citizen science is all about building beneficial relationships and opening up science beyond professional scientists to include a diversity of individuals who may be able to contribute something new to our understanding of the natural world. Volunteers are your partners in this endeavor, not a resource waiting for you to use them. Enter into citizen science as a partnership, and continually keep in mind what you can do to make sure the project is meeting the needs of all partners.

# SEVEN

# Retaining Citizen Scientists

RACHEL GOAD, SUSANNE MASI, and PATI VITT

Citizen science projects vary in length, but many being carried out at present seek to develop long-term understanding. In particular, long-term citizen science projects are needed because ecological questions are not easily answered with one or two years of data. As such, many citizen science projects need not only to recruit participants, but to retain them over time. Thus, retention of participants is of particular importance for many citizen science projects.

Returning participants gain expertise as a result of continued training, educational opportunities, and on-the-ground experience. Such a corps of returning citizen scientists constitutes an informed and expert volunteer group whose contributions are of increasingly high quality. Further, these citizen scientists can serve as volunteer leaders, helping new participants become familiar with the program and extending the reach of program staff. Benefits of long-term involvement are not limited to the program, however. Volunteers who choose to participate over a number of years often find personal value in educational opportunities and deepening involvement in scientific research. While many aspects of citizen science programming affect volunteer retention, the focus of this chapter is on strategies related to program design, training, support and resources, building community, and the meaningful use of data.

In discussing volunteer retention, we are using our own experience with the Plants of Concern (POC) program (box 7.1), a citizen science rare-plant-monitoring program of the Chicago Botanic Garden, as a guide. The POC program was established in 2001 in response to the Chicago Wilderness Biodiversity Recovery Plan (Chicago Region Biodiversity Council 1999) and has since partnered with 120 public and private entities and trained almost a thousand volunteer rare plant monitors to collect data across the Chicago region (figure 7.1). In addition to our own experiences, we have also incorporated thoughts and best practices

FIGURE 7.1. A Plants of Concern field site. Credit: photo by Rachel Goad.

for retaining volunteers from other long-term volunteer programs in the U.S. Midwest and beyond. Although we incorporate some ideas and examples from larger-scale efforts with more distributed systems of volunteer engagement, such as web-based programs with an international scope like eBird, our expertise and network of program contacts are largely focused on local and regional place-based efforts. Hence, while our discussion will be highly relevant to programs with a similar scope of interest, the general principles of long-term volunteer engagement are germane to any long-term citizen science program.

## DESIGNING A PROGRAM THAT SUSTAINS VOLUNTEER INVOLVEMENT

Volunteers are the core of any citizen science project, contributing their time, energy, and intellect to the goal of addressing substantive research questions. Many volunteers in conservation and ecology projects are energized by the promise of meaningful work and the opportunity to make a difference; they should never be perceived simply as free labor. The design of a program should consider the ways in which citizen scientists will contribute and how their contributions will be translated into real-world benefits. The Volunteer Stewardship Network (see box 7.1) has distilled three foundational components for volunteer-based programs: provide meaningful work, empower volunteers to lead, and connect people socially.

Integrating flexibility into data-collection protocols facilitates participation by people with varying levels of expertise, ability, and availability. As such, many programs allow volunteers to fill different roles within a project. For instance, some citizen scientists are

## BOX 7.1   Plants of Concern and the Chicago Area Citizen Science Movement

In the Chicagoland area, a conservation volunteer culture had been developing since 1977, when Stephen Packard and colleagues initiated the North Branch Prairie Project (now North Branch Restoration Project) within the Forest Preserve District of Cook County. Over the next two decades, more such groups formed and were gathered under the umbrella of the Volunteer Stewardship Network, a public-private collaboration of the Illinois Nature Preserves Commission and the Illinois Nature Conservancy. Paralleling this movement, during the 1990s local, state, and federal public agencies, conservation-minded NGOs, and university scientists began partnering to share resources amid regional conservation efforts that culminated in the 1996 establishment of Chicago Wilderness. This dynamic conservation alliance, built from a community of more than five thousand volunteers and many organizations already active in the region (Ross 1997), now numbers almost two hundred institutional and thirteen corporate members from four states. One of the first collaborative efforts undertaken by the members of Chicago Wilderness was the release of the Chicago Wilderness Biodiversity Recovery Plan (Chicago Region Biodiversity Council 1999), a call to action to involve citizens, organizations, and agencies in a broad effort to improve the scientific basis of ecological management and restoration of natural areas, and to develop citizens' awareness of the importance of protecting and conserving local biodiversity. A key recommendation of the Recovery Plan was to establish regional monitoring protocols to support the implementation of evidence-based management of local biodiversity.

Within this conservation matrix and well before the term *citizen science* came into common use, volunteers were already monitoring many organisms and habitats through the Habitat Project (a former project of Audubon Chicago Region and Chicago Wilderness). Monitoring of calling frogs (Peggy Notebaert Nature Museum 2019), breeding birds (Bird Conservation Network 2019), and butterflies (Illinois Butterfly Monitoring Network 2019) engaged numerous citizen scientists. Outside of the Habitat Project network, citizen scientists were monitoring the eastern prairie fringed orchid with the U.S. Fish and Wildlife Service and the Nature Conservancy (Marinelli 2007) and monitoring rivers, forests, and prairies through the State of Illinois' EcoWatch and RiverWatch programs (Latteier 2001).

The region's citizen science program for monitoring rare plants, Plants of Concern (POC), was established in 2001 by the Chicago Botanic Garden, working in conjunction with the Habitat Project of Audubon Chicago Region (now Audubon Great Lakes). From its inception, POC has worked to collect critically needed, region-wide data and to leverage an increasingly engaged pool of ecologically focused volunteers (Ross 1997, Chicago Botanic Garden 2019, Chicago Wilderness 2019). In addition to providing data to local landowners (e.g., county conservation districts) who use it in management planning, POC has also partnered with the State of Illinois to update the statewide database of endangered and threatened plant locations. Tara Kieninger, program director of the Illinois Natural Heritage Database, says, "Because so many groups rely on our data, it is imperative that we keep our database as current as possible. Aside from providing much-needed updates of known populations of rare plants, the POC project gives us information about rare plant populations previously unreported. As such, POC is our largest provider of endangered plant data in the state." POC volunteers are energized by this active partnership and know that their data contribute meaningfully to rare plant conservation in their local area and statewide.

interested in leadership roles while others are interested in participating as an assistant. Other volunteers may be motivated by a communal task or prefer to "work alone and just turn in their datasheet." Citizen scientists with POC may assist an experienced monitor during a field visit, choose to monitor on their own, or can even volunteer in the office, which appeals to those with clerical skills or mobility limitations.

Some programs also engage citizen scientists in work with differing degrees of complexity, as illustrated by Wisconsin's Citizen-based Water Monitoring Network, which offers three levels of monitoring that citizen scientists can engage in (UW Extension 2019). Likewise, citizen scientists with POC have assisted with two levels of monitoring: single-visit population censuses and in-depth plant demographic studies, the latter of which require revisiting the same plants annually to collect detailed data. Another example comes from the Echinacea Project, which is a small but in-depth research program on narrow-leaved coneflower (*Echinacea angustifolia;* figure 7.2) that engages citizen scientists in detailed tasks, like counting and weighing fruits from individual flowering heads (figure 7.3). Citizen scientists are matched with a task they are comfortable with, and they can participate seasonally or when they have time. Flexible work, along with fostering community among the volunteers, communicating results to them, and spending time explaining the purpose of their activities, has created a strongly committed core of participants (Echinacea Project 2019). Building flexibility into a program maximizes participation by meeting individual volunteer needs and comfort levels and serves as a way for less experienced citizen scientists to become more familiar with scientific research and the focus of the project to which they contribute. In place-based programs, the project coordinator can help sort out these personal differences and find an appropriate position within the program for interested volunteers.

Ensuring scientific credibility is an important consideration of program design that is critical for all aspects of long-term program sustainability, including retention of citizen scientists and program partners who value such credibility. One means of safeguarding the scientific integrity of a program is by putting together an advisory group of professional scientists and experts. An advisory group can serve as a key resource, providing feedback on the scientific merit of program directions, helping ensure that all aspects of the program are in alignment with its mission and goals, and weighing in on priorities for future directions. In addition, an advisory group can facilitate partnerships with well-known and respected agencies and organizations that lend additional credibility, support program longevity, and may also lead to funding opportunities.

Lastly, the formation of Citizen Science Central (Cornell Lab of Ornithology 2019) and the establishment of the Citizen Science Association in 2014 (Citizen Science Association 2019) have immeasurably expanded the collaborative capacity of all citizen science programs and may be particularly useful to newly developing programs that are just beginning to consider how to design and implement a project. By offering resources, ideas, support, public recognition, and a community of other citizen science programs, these efforts increase the likelihood of success for long-term programs.

FIGURE 7.2. A bee on a narrow-leaved coneflower.
Credit: photo by Gretel Kiefer.

FIGURE 7.3. A participant dissecting a coneflower.
Credit: photo by Danny Hanson.

## TRAINING

Following recruitment (chapter 6), proper training provides a necessary framework for beginning citizen scientists to understand program goals and their role in advancing them. Training should provide clear, step-by-step instructions for following protocols, as well as the background information necessary to comprehend protocol design. With this solid foundation, citizen scientists will be better equipped to ask good questions, collect high-quality data, solve problems as they occur, and make valuable contributions to the program. Training should also emphasize the important role played by citizen scientists, specifically how their data contribute to larger scientific efforts.

FIGURE 7.4. Participants learning to use GPS. Credit: photo by Rachel Goad.

A variety of training approaches are available to ensure that citizen scientists are well equipped. In-person, hands-on training sessions are an excellent way to connect face-to-face with volunteers, evaluate their needs, introduce them to the project and protocol, and practice necessary skills. After initial training, advanced skill-development workshops can help address specific needs. For example, POC developed an advanced training session focused exclusively on use of GPS (global positioning system) equipment and proper reporting of the resulting data in order to address specific data-quality issues (figure 7.4). Participants have reported feeling significantly more knowledgeable and prepared to collect data after attending this session. Similar training modules can be developed to address needs identified when evaluating project data, and each module should include evaluation of its effectiveness (chapter 12). Connecting volunteers to relevant skill-development opportunities through other organizations can further support their capacity to contribute to the program. Starting and continuing with a solid training foundation will lead to successful performance that helps keep citizen scientists engaged over time.

For volunteers who began with a strong drive to contribute to the project, training and participation may provide sufficient motivation for continued involvement. Others may find more motivation in opportunities that provide transferable skills or helpful tools. For instance, some citizen scientists consider their training and experience with a volunteer program valuable in preparing for a career or avocation in a similar field. A case in point is the set of tools developed by the Cornell Lab of Ornithology for bird watching, including the ability to maintain bird lists collected at different locations across years and how to share these with other birders in a community setting. The development of these tools has significantly increased participation in eBird and other projects at the Cornell Lab of Ornithology (Chu et al. 2012).

## PROVIDING SUPPORT AND RESOURCES FOR CITIZEN SCIENTISTS

As discussed in earlier chapters, one of the most effective means of supporting volunteers is through personalized and frequent contact. While such contact may be untenable in projects with a national or global focus, projects with a more local scope will find maintaining personal connections instrumental in achieving program goals. Individual communication makes it clear to volunteers that they are valued and part of a larger effort, allows staff to respond to a volunteer's needs, and increases the likelihood that volunteers will ask for clarification when needed. Frequent contact with volunteers through meetings, newsletters, group emails, blogs, and social networking posts is a necessary component of a successful, ongoing citizen science–based program. In programs with strongly seasonal activities, like POC, regular communication keeps volunteers engaged throughout the year and not just when their participation is needed.

Individual interaction is essential to the mutual determination of the appropriate role and assignment for each new volunteer, and successful placement is critical to their long-term retention. In-person meetings with volunteers in the field or in the lab (or wherever project data-collection occurs) allow staff to assess a volunteer's capacity, address questions, evaluate the potential for data-quality issues, and gauge whether additional resources are needed. In the case of POC, new monitors are usually paired with someone more experienced who can acquaint them with field protocol. Such support efforts at the outset are critical. Kris Stepenuck, former coordinator of the Wisconsin Water Action Volunteers Stream Monitoring Program, notes that volunteers "can't just be trained and sent out to do the monitoring with no follow-up. The groups that have the highest retention rates not only train volunteers in hands-on training to begin, but the local coordinators go to the field site with each volunteer on their first visit." While individual communication is time-consuming, there is no better method for supporting volunteers, showing them that the effort they make is recognized and appreciated, and confirming that they are appropriately trained and equipped for their task. Additionally, it provides an opportunity to ensure adherence to the project protocols, addressing issues as they arise rather than after the fact.

Volunteer support extends to tools and resources as well. A well-designed website can help support, engage, and retain citizen scientists and could even be considered a critical program element in our digital culture. In addition to housing static resources (e.g., species lists, links, and datasheets), many citizen science project websites are also a portal to the project's database where participants can access and submit data (Sullivan et al. 2014). Websites should be designed to be as clear and easy to use as possible. Difficult-to-use interfaces or frustrating processes on a website will discourage volunteers. In our own situation, the POC website provides password-protected access to past data on populations that volunteers are assigned to monitor. Data can be submitted through the website by volunteers, and a submission triggers emails to relevant partners (e.g., landowners), alerting them to the submission, which they can also access. This alert system allows partners and POC staff to review the data, provide feedback to monitors on data quality and completeness, and clarify protocols that may have been misunderstood or overlooked. Thus, web-based tools can help improve data quality and enhance the volunteer experience.

Lastly, ensuring that all participants can access necessary data-collection equipment (e.g., GPS units and measuring tapes) is an important, if obvious, component of volunteer support. Accessing equipment can be challenging, depending on the regional scope of the project and the type of equipment needed. Equipment funds should be included as a line item in grant or operating budgets whenever possible, to guarantee that enough equipment is available to carry out project objectives. Partnering with agencies and other stakeholders to share equipment can be a solution for extending a project's resources, since similar equipment may be used by multiple projects. Even something as simple as ensuring that all volunteers have the list of equipment necessary to undertake a task can provide an important resource that enhances volunteer preparedness.

## BUILDING A CITIZEN SCIENCE COMMUNITY

Successful long-term programs often cultivate a community of citizen scientists who share a common interest and vision. Such a community is a powerful motivator for volunteers to join up and remain engaged. Newsletters that discuss project results as well as compelling accounts of individual experiences can help communicate a coherent and inspiring vision of program goals and accomplishments. Many programs organize social and recognition events, which foster interaction among citizen scientists and offer encouragement in response to their efforts. Providing refreshments is a time-honored and well-tested way to ensure that people attend and participate in events, and the act of sharing a meal can solidify social bonds. When recognizing citizen scientists, both individual efforts and group accomplishments can be celebrated. Community building is not usually included as a grant deliverable, but it is a recognized benefit of (and to) citizen science projects. Therefore, it is important to include funding needed for events and awards in project budgets whenever possible. However, while this type of recognition is appreciated by many volunteers, it should not be considered a substitute for more substantive responses to volunteer input and needs.

Programs that have a geographically dispersed group of participants may find it difficult for participants to attend in-person recognition events, but many groups find such events to be bond-building and valuable. In the Chicago region, regular joint gatherings involving multiple citizen science programs provide participants a sense of being part of a much larger conservation community, while also reducing the number of separate events that regional citizen science volunteers would have to travel to attend. For example, Wild Things: A Chicago Wilderness Conference for People and Nature attracts both staff and volunteer conservationists who participate in its planning. This long-established biennial conference serves to disseminate information about local conservation and ongoing research and provides a community-building opportunity for the one-thousand-plus participants who attend it. Citizen science programs, their results, and opportunities for new participants to get involved are featured prominently. Where programs are unable to organize such events, they can develop an online presence that facilitates communication among citizen scientists or with staff to ask questions or discuss findings, for example through a discussion board or social media group.

Citizen scientists often work in pairs or teams, and the companionship that develops can foster a sense of community, which is both rewarding and educational. Participants learn from each other, and some find satisfaction in leadership roles by mentoring new members. For example, one POC monitor, having assembled a group of trained assistant monitors, schedules joint field sessions in which a team works together to find and count plants. They consider themselves "buddies with a common bond." In this way, an esprit de corps is forged and mutual learning fostered, building confidence and a sense of accomplishment that contribute to the desire to stay engaged.

A sense of community is also fostered when scientists engage with citizen science volunteers directly to share information or practice data collection. Such opportunities offer interaction and technical support and serve to strengthen the connection between citizen scientists and their work. Group data-collection events that engage volunteer citizen scientists alongside program staff provide an opportunity for citizen scientists to become more confident in their work and more familiar with program staff and other participants. Meetings can be held annually where results are presented to citizen scientists, and during these, time should be allotted for citizen scientists to ask questions and provide feedback on the program. Building these kinds of opportunities into a program encourages the development of an engaged community of committed volunteers.

Listening to and empowering volunteers is a critical factor in cultivating a community of long-term volunteers who can make significant program contributions. Allowing volunteers to take ownership of site-based ecological stewardship work has helped the Forest Preserve District of Cook County's North Branch Restoration Project build a large constituency of committed, knowledgeable volunteers. Likewise, the U.S. Forest Service's Eastern Prairie Fringed Orchid Recovery Team makes a point of including volunteers in decision making on important issues, such as deciding whether to hand pollinate the plants they monitor, and the Calling Frog Survey regularly recruits volunteers to help lead workshop trainings (Chicago Academy of Sciences 2019).

The concept of community can be further extended to volunteer engagement with the natural world, when that is the focus of a citizen science project. The shared cause of caring for the earth—biophilia—can be a powerful motivator for citizen scientists and other volunteers involved in ecology and conservation projects (Wilson 1984). By their nature, volunteers contributing to ecological citizen science care about ecosystems and their components and tend to be passionate about conservation. Many place-based, conservation-focused programs allow citizen scientists to get up-close and personal with their adopted organisms and habitats, follow their progress from year to year, and work to ensure that conservation goals, like preservation of a rare species or eradication of an invasive species, are met. Encouraging this connection to place has been a powerful motivator for volunteer involvement in citizen science programs of the Chicago region.

## THE ROLE OF DATA IN SUPPORTING LONG-TERM PROJECTS

Citizen science data that are effectively managed and utilized can energize volunteers, as they see their contributions leading to meaningful results (Kelling 2012). Effective data use also motivates partners to continue their involvement because access to a useful, well-curated, long-term dataset becomes invaluable. Not surprisingly, funders are more likely to continue their support when presented with clear and useful results.

Establishing means by which results of data analysis are delivered to participants is an important component of any program. The eBird project provides a high-tech example of how data delivery can be done, with results translated into interactive maps and graphics (https://ebird.org). Likewise, the Eastern Prairie Fringed Orchid Recovery Team holds annual meetings at which researchers share results from analysis of the data collected by volunteers. This group and others have found that volunteers who learn about the project they are contributing to are motivated to maintain their involvement. In the case of POC, we communicate program results through meetings, newsletters, outreach pieces, website content, and published research, while keeping volunteer motivation in mind (Fant et al. 2008, Vitt et al. 2009, Bernardo et al. 2018).

More than simply reading about program results, citizen scientists who see the data they have collected having an impact in real and measurable ways gain a tremendous sense of satisfaction and may be more motivated to continue participating. POC data influence land management on a very local scale, addressing acute threats like mowing of endangered species or invasion by non-native species. When landowners address such threats in response to data submitted by citizen scientists, participants may see on-the-ground changes in the populations they monitor and feel empowered by their participation in the program.

Citizen scientists should also have the opportunity to provide feedback about their experience collecting data. Since participants serve as a program's "boots on the ground," many become experts in data gathering and have significant insight to offer. Citizen scientists may be the first to notice that a protocol needs to be adjusted to account for real-world conditions, or their feedback may indicate that additional training or resources are needed. Affording such opportunities is critical to both clarifying protocols and procedures and retaining active participants.

## SUMMARY

Citizen scientists generously commit their time, skill, and passion to a scientific initiative in the belief that they can make a substantial and meaningful contribution. However, effective engagement of these participants requires careful planning and implementation. Be sure to show volunteers that their efforts are recognized and their contributions utilized, and remind them that they are collaborating with a focused community of like-minded participants. While the suggestions and points discussed in this chapter are not exhaustive, they do represent well-tested methods of long-term volunteer engagement, particularly in place-based citizen science. Thus, we suggest the following best practices as a useful guide in developing a citizen science project:

Communicate an inspiring vision of the work to be done.

Set clear scientific goals for project data and maintain scientific oversight to establish program credibility for volunteers as well as partners.

Provide frequent communication and, when possible, individualized attention. Be vigilant in balancing administrative demands and volunteer communication.

Provide flexibility and adaptability to volunteer needs.

Accommodate a variety of backgrounds and learning styles while simultaneously communicating the need to follow scientific protocol in a rigorous manner.

Acknowledge citizen scientists as partners in a common cause and solicit their input.

Do not belabor the idea that citizen scientists save professionals money, and be wary of phrases like "using citizen scientists."

Provide results and feedback on what the data show, how they are used by scientists, and how data collected by citizen scientists make a significant contribution.

Disseminate results and demonstrate program credibility through publications, press releases, and an engaging website.

Provide ongoing training and learning opportunities for citizen scientists.

Develop volunteer leaders by involving experienced citizen scientists in training others and serving as advocates for the program.

Nurture camaraderie and social connections to foster community.

Encourage a "connection to place" for volunteers working in place-based programs.

Give meaningful appreciation and recognition.

Prioritize long-term sustainability when designing a program, and consider how funding, partnership, and institutional affiliation can help support that goal.

# EIGHT

# Training

HEIDI L. BALLARD and EMILY M. HARRIS

Most citizen science projects require some type of training for participants (as discussed in chapter 7). Training can provide instruction for proper data collection and serve as a communication tool between project leaders and participants. Whether in person or online, training can be one of the best opportunities for project leaders to inform participants of the scientific, conservation, and educational goals of the program. In turn, a training session may be the first (or only) chance for participants to communicate their interests and expectations to project leaders. Therefore, careful planning, implementation, and evaluation of citizen science training can support many of the goals of a citizen science project. In this chapter, we introduce some of the different goals for citizen science training and effective ways to accomplish them. We discuss types of training, give some brief background on what educational researchers understand about how people learn, and offer detailed suggestions for facilitators[1] who wish to provide structured, effective sessions on collecting, submitting, and analyzing high-quality data that are engaging, informative, and a rich experience for participants.

1. We describe the people developing and/or implementing a training as facilitators (whether they are scientists, project leaders, educators, or volunteers) to emphasize that the facilitator's role is to *facilitate* hands-on learning experiences for participants to learn, practice, and get feedback, rather than solely lecture to participants.

## GOALS OF TRAINING

### Quality Control and Quality Assurance

Training is one of the key methods for delivering quality control and quality assurance to citizen science programs. In particular, training can evaluate, provide feedback on, and check the performance of citizen science participants in relation to data and other protocols. Furthermore, thoughtful and thorough training helps ensure that participants understand the underlying goals and rationale for the way they will be collecting data or accomplishing other parts of the project, which means they will be conscientious and thoughtful contributors rather than simply technicians.

### Broader Science and Environmental Learning

Citizen science project leaders often make a range of assumptions about what their participants will learn through participating in their project. These assumptions include a better understanding of the scientific process and developing a stronger environmental stewardship or conservation ethic. Fortunately, educational research is emerging to test these assumptions. A key question is which of these various learning outcomes result from the activities of data collection and analysis that participants engage in, and which of them result from the training program. Depending on the structure and nature of the citizen science project, the training sessions could be the best and last chance that project leaders (as facilitators) have to facilitate the learning goals for their participants. For example, if your goal is for participants to understand the connection between monitoring birds or plants in their backyard and their own behaviors of spraying pesticides in their gardens, you must be explicit about the behaviors you hope to influence (Monroe 2003), and the training session is an ideal time to explain this clearly.

Many facilitators assume that participants will learn through exposure to the larger conceptual or behavioral connections between the skills and tasks of the project and the bigger picture of science and the environment. However, exposure is usually not enough to effect such learning. By providing the background, context, and history of the project during training, facilitators can be confident that they are targeting the key things they hope participants will learn (chapter 12 describes how to evaluate whether you have met your learning goals for the project overall). Finally, a key learning goal of citizen science training is to facilitate participants' meaningful engagement with a project from the beginning in a way that speaks to their own personal motivations and interests. Thus, incorporating participant motivations is an important part of training. One way to incorporate motivations is by taking time at the start of a training to find out what motivated the participants to come in the first place, and what they hope to get out of the project. These questions can be answered either through discussion or with a more formal pretraining questionnaire, which facilitators can use first to guide their approach to the training and then to frame the rest of the project for participants in ways that resonate with their motivations.

## TYPES OF TRAINING

Citizen science training is carried out in a variety of contexts, settings, and locations, with different audiences and organizational partners, and through many different mechanisms. That said, training typically involves two main types of approaches: in-person or online. Depending on the nature of your project, you may decide to use either or both of these approaches.

### In-Person Training

Citizen science projects often have two stages of in-person training: initiation or orientation training, and refresher training. During the former, participants learn project goals, content and background, tasks they will be doing, and new skills that they will need to use, including data entry in the field or when they return home to a computer. This initial training often occurs before the trainees have committed to ongoing participation and thus may include an element of recruitment and "selling" the program. Aside from initial training, some projects include ongoing training or refresher sessions for existing participants. This can ensure they keep their project skills sharp and can provide a sense of community and a social dimension, particularly if participants do not typically get to see each other often to compare notes about what they are doing in the field. Longer-term projects that have long time spans between field seasons (e.g., migratory bird and butterfly monitoring or plant phenology projects) should consider providing these refresher training sessions to ensure that participants remember protocols and identification strategies.

### In-Depth Training for Long-Term Projects

In-person training can take many forms, depending on the structure of the project. For long-term projects, the general rule is that it is always better to conduct a training in the habitat in which participants will be collecting data. That being said, it may not be possible to train at the field site, or there are practical reasons to hold the training indoors where participants can better focus on the information before being asked to learn and perform tasks. However, indoor training should still include hands-on activities as much as possible, such as practicing identification of specimens or use of special tools like calipers and hand lenses, so that participants can become familiar with materials and procedures.

### BioBlitzes and One-Time Citizen Science Events

In contrast to long-term projects, some citizen science programs are one-time events, such as BioBlitzes or meet-up events. Training for these short events will likely be quick and take place on site where data will be collected. Natural history museums, science centers, municipal governments, Cooperative Extension programs, and community-based organizations are increasingly offering these one-time events to entice participants who might not be ready to join long-term projects. For one-time programs, we suggest you use the same principles and guidelines discussed here for the more in-depth training, but condense the training time to be proportional to the amount of time people will be participating in data collection overall. For example, if the BioBlitz event is to last six hours, the training should probably take thirty minutes to an hour. Another common feature of BioBlitzes and other one-time events is a train-the-trainer model in which past project participants are enlisted to help train and perhaps lead groups of new participants during the event. In this case, it is important to provide those participant leaders with training in advance of the event (the morning before or earlier) so they can feel confident and be prepared.

### Online Training

Online tools provide a wealth of resources for citizen science project training. In fact, many projects now include online training materials and online data submission. Moreover, much of the communication that occurs between the project leaders and participants happens online, depending on the audience. As such, you should make sure to set aside time in your training sessions for participants to learn and practice the online tasks that are required of them. However, it is important to remember that not all participants will have access to or feel comfortable with all online tools, and hence it is important to check in with them and provide, if possible, paper versions of surveys, data protocols, newsletters, and so on.

Many large national and international citizen science projects, such as the USA National Phenology Network (www.usanpn.org), NestWatch (https://nestwatch.org), the Monarch Larva Monitoring Project (https://monarchlab.org/mlmp), and the Great Sunflower Project (www.greatsunflower.org), provide all the training and information needed to participate through online resources on their websites. However, many national and international projects are easily adapted by locally or regionally focused programs that organize training and data collection in a specific place (e.g., park, preserve, or neighborhood) and contribute their data to the larger project. Using such existing projects can be an efficient and useful way to engage more people in citizen science without having to reinvent the wheel and design an entire program from scratch, especially if a given program meets your goals for scientific data and participant engagement and learning. Local project leaders can create a constituency of committed participants who will monitor a park or neighborhood, by offering in-person training to complement national projects like those mentioned above.

Finally, many project leaders rely on the Internet to communicate with participants about training, through newsletters, email, and social media. A variety of online tools are available for communicating with participants about training, including those found at CitSci.org, which has templates for newsletters, advertisements, and so on. In addition, you may want to use online survey tools such as SurveyMonkey (www.surveymonkey.com) to conduct surveys with participants both before and after the training. Surveying beforehand is useful to gauge the backgrounds and interests of participants, and surveying afterward is an easy way to evaluate the effectiveness of the training session.

## ELEMENTS OF TEACHING AND LEARNING RELEVANT FOR TRAINING

Depending on whether your audience consists of youths, adults, or both, developing and implementing training is conceptually somewhere between a teacher developing and teaching a lesson plan and a facilitator running a workshop. While youth and adult audiences are different in some key ways with respect to citizen science training (as discussed below), many of the principles of good teaching and facilitation are the same. In particular, education and environmental education research indicate the following principles about how people learn (paraphrased from Wals and Dillon 2013):

Learning requires the active, constructive involvement of the learner.

Learning is primarily a social activity.

People learn best when they participate in activities that are perceived to be useful in real life and that are culturally relevant.

New knowledge is constructed on the basis of what is already understood and believed.

Learning is better when material is organized around general principles and explanations, rather than based on memorization of isolated facts and procedures.

Learning is a complex cognitive activity that cannot be rushed. Considerable time and periods of practice are required to build expertise in an area.

Learning is critically influenced by learner motivation, so consider what interests and needs might bring someone to a training.

### Training Considerations for Adults

Youth and adult audiences likely come to citizen science training sessions with different interests and expectations, which will influence the design and implementation of the training. We know that adult learners differ from young learners in some fundamental ways. Specifically, adults often have vast experiences and a wealth of knowledge that they bring to any new educational experience, and they typically respond well when their knowledge and experience are respected and utilized within the training (Jacobson et al. 2016). And, while all learners need to see the relevance of what they are learning to their own lives, adults particularly need to be able to apply and reflect on what they are learning and why it is important in the bigger picture of the citizen science project. Finally, it is crucial to remember that adults are volunteering their time of their own free will, so if a training activity does not seem useful or meaningful, they can simply walk out. Therefore, make sure to communicate clearly what the training will entail and why, so that people have clear expectations when they arrive.

### Training Considerations for Children and Youths

Young people may or may not be participating in a project on the direction of a teacher, a parent, or another adult. To support young people's intrinsic motivation, emphasize the real scientific audience, create opportunities to recognize their contributions, and support their social interests. Some young people are motivated because they feel that citizen science projects offer an opportunity to participate in real science (Ballard et. al. 2017) and/or connect with and steward a place that is important to them (Haywood 2014). Furthermore, recognition of young people supports their ability to see themselves as science experts and can include receiving badges or recognizing specific achievements, such as who registered the most observations. Providing food at training is appreciated by youths and adults alike. Family and other social dynamics can also influence how well participants engage with training activities. In some cases, it makes sense to separate families from other adults for a few skill-building activities that target each generation's interest and ability levels. However, if young people are attending a training with adults, they will likely be collecting data with those adults, so it is also important to design activities that are accessible and interesting to both audiences simultaneously, and allow them to work together as a team (see some suggestions below).

Citizen science programs are increasingly used as a tool for teachers to engage young people in science in formal school settings (Harris and Ballard 2018). Several large, national citizen science projects—such as Driven to Discover (https://extension.umn.edu/citizen-science/driven-discover) and BirdSleuth (www.birds.cornell.edu/citscitoolkit/projects/clo/birdsleuth)—have developed science-focused curricula that allow teachers in schools to train their own students to conduct data collection, which not only helps children learn scientific concepts and processes but can also contribute high-quality data to the citizen science project. In other cases—such as LiMPETS (https://limpets.org) and Vital Signs (http://vital-signs.org)—teachers and after-school educators can attend training workshops in their local area and then conduct field-site data collection with their students that contributes to a larger dataset used by scientists. In both cases, the project leaders need to target training materials to the teachers, parents, or educators who will then work with the students themselves. Thus, if you are working with teachers, you may need to design background material for them as well as more detailed behind-the-scenes instructions, in addition to the standard protocols that the youth and adult data collectors will use (see the "Youth-Focused Community and Citizen Science" resources at http://yccs.ucdavis.edu).

## HOW TO PLAN, IMPLEMENT, AND EVALUATE A TRAINING

### Planning

#### *Big-Picture Planning*

Establishing the goals and desired results (Wiggins et al. 1998) of the training is an important first step of the planning process. In determining your training goals, consider your participants (see chapter 4), the anticipated timeframe for their involvement (e.g., one day or ongoing), and what parts of the scientific research they will be involved in. In addition to accurate data collection, consider other goals you may have, such as participants being able to (1) analyze and classify data, disseminate conclusions, or engage in other parts of the scientific research; (2) understand why the research questions of the project are important and how they are contributing; (3) teach members of the public about the project while they are in the field; and (4) understand the nature of scientific knowledge generation and "what scientists do."

Once you have established training goals, determine what evidence is needed to demonstrate that participants have met the goals (Wiggins et al. 1998). That is, how will you know that the participants understand and can do what you had hoped by the end of the training? One way to assess participant understanding is to have them perform a task—an activity in which they do something similar to what they will later be expected to do on their own during data collection or other project activities (often called "performance-based assessment"). For example, if a desired goal is for participants to correctly identify an organism with 80 percent accuracy while monitoring with a partner on a field transect, an appropriate performance task would be to have participants spend fifteen minutes at the end of the training session working in pairs to identify twenty organisms. Similarly, if training for species identification or classification of photos occurs online, make sure that participants get to practice classifying and testing themselves before doing it for real. For more detail about evaluating whether participants have learned what you expected, see chapter 12. Starting with these goals and the tasks or evidence you will use to determine whether the training was successful, you can use them to "backward design" a plan for a training session of appropriate learning activities that meet the training goals (box 8.1).

### Planning the Learning Experiences

When planning the details of the training, project leaders and facilitators should include a variety of activities and teaching styles beyond just direct instruction (i.e., one-way verbal presentation of information). Depending on your training goals, these activities can include presentations, investigations of the species or phenomenon of interest, time to practice protocols with feedback from instructors, small- or whole-group discussion, and participant reflection about what they have learned and how to apply it. Some activities will be most appropriate as presentations, while others will be more effective in a hands-on format. You should strive for a balance in time between how much you will be talking in a presentation and how much participants will be actively engaging in hands-on activities. If building a social community among participants is a goal, include activities that facilitate participants connecting with each other. In order to determine what needs to be included in the training and what kind of resources will support participants in the field, consider their expected roles in the project. For example, how often do you expect participants to engage in monitoring, are they monitoring alone or in groups, and what kind of data will they collect?

### Interactive Presentation

Presentations interspersed with audience participation can be effective for developing conceptual knowledge, drawing out participants' knowledge and experience, sharing the larger context in which participants are working on the project, introducing participants to specific parts of the work, or making meaning of what participants are observing. For example, at the start of a training session, many facilitators lay out the larger context of the research, including what

12:00–12:45 Lunch

12:45–2:00 Practice the monitoring protocol

Group plant practice and data sheet (whole-group discussion and modeling the monitoring protocol)

Observation practice in the arboretum in small groups/pairs (small-group practice using the protocol with several individual plants)

2:00–2:15 Break

2:15–2:45 Uploading the data

Orient participants to data entry online (presentation)

Participants enter data together in small groups (guided practice)

2:45–3:00 Closing

Closing activity

Post-training survey

*Day 2, in the field at the location of the trail*

9:30–10:00 Introduction

Icebreaker

Plan for the day

10:00–10:15 On-site materials

Orientation to different trails (hand out maps)

Demonstrate lock box and proper storage of data sheets

10:15–10:45 Whole-group plant monitoring

Participants practice together on one plant and then have whole-group discussion

10:45–12:00 Small-group monitoring

In groups of three or four, participants practice monitoring while facilitators circulate to answer questions and provide feedback

12:00–12:30 Elevator pitches

Develop elevator pitches

Practice giving them to other participants

12:30–1:00 Lunch and closing circle ("Tell the group one thing that surprised you today")

research questions the scientists are asking, what progress has been made on the research, and why the research is important to the scientists and to the public. This type of presentation could happen as either an in-person training or a module in an online training.

Presentations do not need to follow a one-way form of communication. Instead, presentations are better if they are interactive and engage participants throughout. Telling specific, compelling stories during the presentation—in particular, stories from the professional

scientist's perspective of curiosity and discovery or from a prior participant's experience—
can inspire and motivate new participants. In addition, eliciting participants' interests,
goals, and prior knowledge not only breaks up a presentation, but also helps instructors better understand participants' motivations and levels of existing knowledge. Including questions for discussion, such as "think-pair-share" (think about a question, turn to your neigh-
bor, and discuss it for two minutes) or thought exercises (where learners can think through
an issue or question before the whole group discusses), can also help engage participants,
particularly language learners or people who do not like to talk in large group settings.
Finally, checking for understanding and using verbal or nonverbal cues not only keeps participants engaged, but can also indicate whether participants are ready to move on or need
more time to discuss and review (for examples, see box 8.2). When planning a presentation,
always ask yourself if there is an interactive way to engage participants. Also try to include
more than one presenter or facilitator, if staffing allows, which varies the energy and
approach to the topic and keeps people engaged.

## Hands-On Activities

Hands-on activities provide an opportunity to teach many different skills, including organism identification, data-collection methodology (e.g., how to estimate or accurately measure), following a protocol, and how to upload data. In many cases, such hands-on activities can be derived directly from a presentation (box 8.3).

One of the most important hands-on parts of a training session is teaching new participants to accurately follow a protocol. For citizen science projects in conservation and ecology, where species identification or knowledge is a key component, it can be valuable to divide the training for the protocol into the following three parts (building on Jordan et al. 2012; their related cognitive tasks are shown in parentheses).

### Identify and Recall Species of Interest (Object Identification and Memorization)

Instead of describing different structures of a species in a presentation and then asking participants to immediately identify a specimen, participants can first compare and contrast specimens by sorting them. This sorting allows participants to construct their own understanding of important structures that will help them identify species (Brooks and Brooks 1993). A small tweak in how the learning is structured will aid in identification and memorization of important differences (box 8.3).

### Detect the Species of Interest in the Field (Discrimination and Authority)

Detecting a species in the field is more challenging than identifying it in a classroom or lab setting. Thus, have participants practice identifying specimens in an environment where the organism is obvious and then gradually introduce more complex situations. This obvious-to-complex pathway of learning allows participants to develop a language for and understanding of the species. For example, if participants are learning to identify birds, they should start by observing key features of a species (e.g., field marks) by looking at pictures and specimens in the classroom. Then, when the participants are in the field and a bird flies by, they already have some level of familiarity with the key features they are looking for on the bird.

### Appropriately Follow the Data-Collection Protocol (Accuracy and Automaticity)

In order to accurately collect data for the project—which will require a level of expertise and automaticity in enacting the protocol correctly—participants need opportunities to practice the protocol (Monroe 2003) with support from the trainers. They should practice in small groups at first, and then either alone, in pairs, or in groups—however they will be working during actual data collection. One way to accomplish protocol training is by gradually shifting the responsibility for the activities from the facilitator to the participant (Collins et al. 1989; figure 8.1). For example, the facilitators start by modeling how a protocol should be run for the entire group of participants, which allows participants to become familiar with data-collection sheets or apps and develop a conceptual model of the whole process. Then the participants break into groups of three to six to practice the protocol together, with the

BOX 8.3   From PowerPoint to Participation: The California Phenology Project (CPP) at Stebbins Cold Canyon Natural Reserve

Here is one example of how to transform a presentation-style session into an active hands-on learning activity. For our CPP Stebbins project, our team had outlined a well-developed forty-five-minute botany PowerPoint presentation that introduced the six plants participants would be monitoring and important botanical structures. This presentation was scheduled to happen after another forty-five-minute introductory PowerPoint introducing what phenology is, the link between phenology and climate change, and the places where our project participants would be fitting in. However, we were concerned that ninety minutes of lecture would leave participants sitting for too long, so we decided to modify the plant presentation to be interactive, to help participants identify different plant structures in a hands-on way.

WHOLE-GROUP OBSERVATION PRACTICE

We decided that having participants explore the six plants in small stations would allow them to construct their own knowledge and understanding of each plant. Not only that, but in the process of observation they would notice many important structures that we could reference afterward in our botany lesson. Because our training was not close to the monitoring site, we decided to bring cuttings of plants for the participants to explore. We introduced the activity by stating the goal of the session: to introduce participants to the six plants they would be monitoring. We asked participants to make a chart in their notebooks with two columns: *I notice* and *I wonder.* Then we passed out a cutting of one of the plants to each of several small group of two to four participants and had them record statements about things they noticed about the plant and questions they had. We also had them come up with a name for the plant based on characteristics they had observed. We emphasized that we did not care if people knew the name of the plant and encouraged them to make careful observations. After about five minutes of exploration, we brought all the small groups together and they shared their observations and plant names. In this whole-group debrief, we encouraged participants to expand on their observations with prompts such as these: Can you explain what you mean by _____? What did you notice that made you wonder that? Why did you name your plant that? This led to participants articulating their observations of the characteristics of plant parts.

SMALL-GROUP OBSERVATION PRACTICE AND FORMATIVE ASSESSMENT

Once everyone had practiced together, the small groups rotated through the other five stations, doing the same activity for each of the other five plants. After visiting all the stations, we had a whole-group discussion, sharing notable observations for each species and pointing out specific plant-part structures to pay attention to. To conclude, we put out six species profile sheets that had images of each plant with all of their phenophases as well as the name of each plant and additional information about them. We had the small groups work together to decide which of the species they thought each of their specimens was and gave them time to read about each plant on their own. After getting to know the plants so in-depth, participants were motivated to read about the plants and felt proud of the ease with which they could identify them. This was an exciting way for participants to get to know the six species they would be monitoring, and their close observations helped them see many of the botanical structures we had planned to show them in the presentation. Later in the day, we reviewed one slide on basic botany and were able to reference specific plants that participants had observed, which helped ground their learning. In this way, we turned a forty-five-minute lecture into a forty-minute activity with five minutes of follow-up PowerPoint slides.

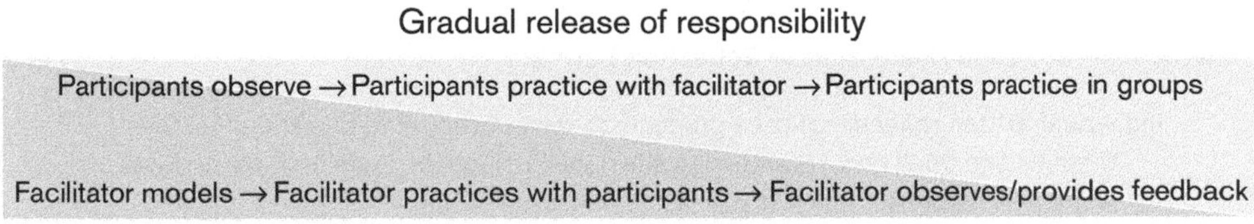

## Gradual release of responsibility

Participants observe → Participants practice with facilitator → Participants practice in groups

Facilitator models → Facilitator practices with participants → Facilitator observes/provides feedback

FIGURE 8.1. Model of gradual release of responsibility by a facilitator, whereby the participant's role shifts from observer to practicing the skill in groups, and the facilitator's role shifts from modeling the skill to observing and providing feedback.

facilitators working alongside them to answer any questions. Finally, the participants do the monitoring activities themselves (singly or in teams as the protocol specifies) while the facilitators watch and offer feedback as needed. Notably, this gradual release of responsibility is not always the best approach. For example, if the cognitive demand of the task is minimal, participants can move immediately into self-directed activities without a facilitator first modeling how to do it, as in the aforementioned sorting of specimens to create an observation-based conceptual model they can draw on later in classifying organisms.

Other activities that might benefit from hands-on training include uploading data and sharing information about the project with passersby while working in the field. In these two situations, you should model what the activity looks like and then give participants the chance to practice in small groups or pairs. If participants need to learn how to navigate the project website and upload data, you might model how to do it and then have participants practice together while you circulate to help them. Likewise, if participants will be expected to share information about the project with passersby, you should give them a chance to practice giving a short explanation about the project to another participant or an actual passerby.

### Group Discussion and Reflection

Opportunities to talk about and reflect on new content, skills, or processes are critical for developing meaning about them. Reflection allows participants to monitor their own learning and recognize if they need extra support or more practice (Schraw et al. 2006). Discussion and reflection also help identify places where participants are struggling so that you can adjust the training to meet their needs. There are many ways to facilitate this, including individual written reflection, partner conversations, and small- and whole-group discussions (box 8.4). These strategies should be incorporated throughout the training—in conjunction with a presentation, during hands-on activities, or as stand-alone exercises.

### PLANNING THE LOGISTICS

As you develop your detailed plan for training activities, consider what we know about what motivates people to learn. Motivation is often described in psychology as an internal desire

**Individual written reflection.** Post a prompt you want participants to respond to. Give them time to think about it and write down their responses. Then have participants share their responses either in pairs or to the whole group.

**Small-group discussion.** Small-group discussion allows everyone to talk and process, and some people are more comfortable sharing their thoughts with one or two people than with a whole group. In a practice setting, give each group one or two prompts to discuss. Allow time for participants to talk with each other while you circulate and sit in on a few conversations. Have each group select one person to report to the whole group on their group's conversation. This approach can be useful when participants are practicing skills, such as classifying organisms or estimating.

**Whole-group discussion.** This can be a great format for the opening and closing activities of a training. You can give everyone the chance to speak to the whole group about what brought them to the project or what they are excited about moving forward. You can also use this strategy to encourage group discourse, which is particularly useful when practicing protocols. Encourage participants to share their thinking processes out loud and comment and respond to each other. Act as a facilitator more than a participant to unearth the group's ideas. Then you can step in and provide feedback on the ways people are thinking about the monitoring.

or want that activates or energizes behavior and gives it direction. Taking into account what might motivate participants in a citizen science training could dramatically enhance its effectiveness for learning. Consider the following logistical items, many of which come from our understanding of factors that make people comfortable and motivated to learn. Keep in mind the importance of meeting people's physical needs first, before addressing their cognitive needs (Maslow 1954).

### Location and Space

People need to have their basic needs met in order to be able to focus, so finding the right location and space is crucial for a smooth training session. Begin by considering the following questions:

What is your anticipated capacity? Make sure you have space to accommodate everyone comfortably.

Will your training be indoors, outdoors, or both?

If the training is outdoors:

Are there places where your whole group could stand in a circle?

Are there benches or other places where participants could sit if they need to?

Are there places that everyone in your group will be able to gather around to look at examples?

Are there shady or sunny places to help participants feel comfortable?

If the training is indoors:

Will participants be sitting? Lecture-style? In small groups with tables? What setup will facilitate participants working together?

Is there a nearby outdoor location where participants can practice?

Are there any site-specific considerations? Is there enough parking? Are there restroom facilities?

Considering these points will ensure not only that participants are physically, emotionally, and psychologically comfortable during the training, but that they are engaged and excited to participate.

### Staff Help

During your planning, be mindful of the facilitator-to-participant ratio and incorporate this into your activity design. If you have a lot of staff, you can divide into small groups. However, if you do not have a lot of staff, consider mixing people into groups with a range of experience levels so that they can help one another. Consider having experienced participants help facilitate or take the lead (with staff support) to run or even plan training—they can be great assets, with extensive experience and local project knowledge.

### Materials and Gear

Determine and document all the materials you will need for each training activity (box 8.5). Consider the size of groups that participants will be practicing in and make sure you have enough materials for your intended design. You should ask yourself several questions *for each activity* to ensure that the materials you have are useful and sufficient:

Will participants be working in groups? How many will be in each group? Will they share materials?

Do participants need to write down information? If so, on what will they write? If in the field, will they need clipboards?

Do participants need hard-copy instructions in addition to oral instructions and modeling?

### Safety Management

Prior to the training, identify important hazards and possible risks for your group. These might include, but are not limited to, sunburn, dehydration, cuts or scrapes, hypothermia, biting insects or other invertebrates (e.g., ticks), animals to be aware of (e.g., snakes and large mammals), poisonous plants (e.g., poison oak and poison ivy), and personal medical health problems. Come prepared with extra supplies for participants who forget or can't afford the typical gear, and a well-stocked first aid kit. Plan for emergencies so that if something unexpected happens you will be ready for it. Locate the nearest hospital or medical facility and determine whether you will have cell phone reception if you are at an outdoor location.

### Communication with Participants Prior to Training

Communication before the training session will help everyone have clear expectations. Along similar lines as participant recruitment (chapter 6), advertise early and often about the training, through a variety of formats, including fliers in key places, announcements on the project website and the websites of related organizations, radio, newspaper ads, and list-servs. Make sure to emphasize the importance of good training to the science and success of the project, and that the training will be fun and engaging. You will likely want to send several reminders to participants, including one reminder a week before and a final reminder one or two days before the training. Your communication with participants should include the following:

What to bring (clothes, materials such as cell phones or writing implements, food)

Location, with explicit directions for how to get there and parking information

Agenda for the day

Whether you will be providing food and/or beverages

Cell phone number for your team in case people need to contact you the day of the training

Finally, box 8.6 contains an example of a detailed plan—for an activity called "Meet the Plants," based on the activity we described in box 8.3—including the objectives, the materials needed, and the steps the facilitator will follow to do the activity as part of a training. Documenting your training activities like this can be crucial if multiple staff members will be conducting training sessions for the project or if it is a recurring event.

### ONGOING TRAINING

Some citizen science programs have documented a first-year or learning effect (Jiguet 2009). That is, participants collect more accurate data over time as they become increasingly familiar with the observation protocols, more adept at identifying the species of interest, and more aware of where and when species will occupy certain areas. This effect occurs not only with citizen scientists but with professional scientists participating in monitoring programs (e.g., breeding bird surveys; Sauer et al. 1994). Consider planning follow-up refresher sessions with participants soon after they start monitoring on their own so that they feel confident in their abilities and so that you can intervene if necessary before participants develop habits that are difficult to reverse.

The data-collection needs for some projects change over the course of a season or year. If such changes occur, plan training sessions to coincide with the changing environmental conditions. To aid in how trainings change over time, it is useful to map out expected data-collection protocols throughout the year when determining when to hold your training and follow-up sessions. For example, in winter it is difficult to distinguish between phenophases (progressive stages of leaf emergence, flowering, fruiting, and leaf drop) on deciduous perennial plants. In this case, the ideal time to host an initial training session is in the springtime, with follow-up sessions timed throughout the summer and fall as new phenophases emerge. Alternatively, you can share resources, such as photos and videos, and host discussion forums in which participants share their observations and discuss changes in their observations.

BOX 8.6   Detailed Activity Plan for "Meet the Plants"

OBJECTIVES

Participants will be able to:

make and document initial observations of six focal species,

identify the six focal species, and

develop their own questions about the six focal species.

MATERIALS

Cuttings of all six focal species (four or five of each species, enough for small groups of four or five participants; even better if there are cuttings from different phenophases).

Notebook and writing implement (one per participant).

CPP Stebbins species profile sheets (one of each plant).

ACTIVITIES

*Introduction* (five minutes)

Divide the whole group into five groups total, with up to four or five participants in each group.

Introduce the goal of the activity: This activity will introduce you to the six focal species we will be monitoring at Stebbins. If you do or don't know what these plants are, *that's OK!* We will practice making careful and close observations with the goal of getting to know the specific plants and documenting observations about them.

*Whole-group practice* (ten minutes)

Have participants make a "T-chart" in their notebooks with column headings *I notice / I wonder* and include a Plant Name title at the top of the page. (Do not fill out the Plant Name title for them.)

Plant Name:
I notice                          I wonder

Give each group a cutting from the same plant. Have them work together to fill out their "I notice / I wonder" T-chart. Prompt participants to make many observations about all different parts of the plant structures. If they don't

## IN-THE-MOMENT FACILITATION TECHNIQUES DURING TRAINING

Even after meticulous planning, effective training often comes down to in-the-moment strategies that facilitators use to keep participants engaged, reinforce important concepts, incorporate lived experiences, and include multiple perspectives that enrich the entire experience and ultimately accomplish the goals of the training. Keeping in mind some key know what a particular structure is called, encourage them not to worry about it.

Have each group develop a name they would call the plant. If time permits, have them make a quick sketch of the plant.

*Whole-group share-out:* Have groups share their observations and plant names. Reinforce observations and have some participants also share their "I wonder" questions. Encourage asking questions.

Prompt participants to explain their observations, questions, and creative/ memorable names. Example questions:

Can you explain what you mean by_____?

What did you notice that made you wonder that?

Why did you name your plant that?

*Small-group practice* (ten minutes, ideally at least five minutes per species for a total of twenty-five minutes)

Have each group make another "I notice / I wonder" T-chart.

Have groups disperse to five different stations, each with cuttings from one of the five species of plants we are monitoring.

Have them fill out the chart: making observations and developing questions, recording their observations, and coming up with a name for each plant.

Allow groups to rotate freely after they have completed a species. Have them use a separate page and chart for each plant.

(There may not be enough time for each group to get to all the species.)

*Whole-group discussion* (ten minutes)

Bring the group back together.

Go through each species, having the small groups report to the whole group their observations and the creative/informative names they created for the different species.

*Assessment* (ten minutes)

Give each group one of the species profile sheets.

Have them work together to decide which of the species they think it is.

Have them place the species profile sheets next to the cuttings.

Have a gallery walk in which participants can go around to each cutting and look at the species profile for each of the plants they observed. Participants may want to add notes to each of their observation pages.

techniques throughout the training sessions can help achieve these objectives. Such techniques include the following:

Allowing enough time for every participant to try every activity or task, possibly dividing the group into smaller groups so that more activities can be hands-on.

Providing opportunities for participants to figure out the answer to a question or to do a task themselves, rather than telling them or doing it for them.

Giving lots of critical, but supportive, feedback during and after participants practice new skills, especially in the field, so that they know when they are doing things accurately or correctly.

Alternating between inductive and deductive processes. That is, sometimes allow participants to study a phenomenon or make observations before you provide the content information (like plant or insect characters used to identify species), and sometimes provide the information first and then let them explore and observe.

Checking for participants' understanding frequently throughout the training, verbally and nonverbally (i.e., asking people to perform tasks, not just answer questions).

Perhaps most importantly, learning how to "think like a scientist"—from classifying organisms to estimation—is difficult to observe and can be explicitly scaffolded (providing tools, examples, and practice) to help participants become comfortable with monitoring protocols (Collins et al. 1989). When stepping in to assist participants with a data-collection protocol, facilitators should model their thinking out loud while figuring out a tricky specimen identification or analyzing findings. This assistance helps participants learn how they should be approaching the task at hand and identify where they may have strayed in their thinking. As participants become more experienced through practice, you can ask them to talk out loud about their thought process and then provide feedback to help them adjust their thinking. If the training is for a larger number of people (more than twenty), most of these techniques can be done by breaking the group into small groups of four or five, providing clear instructions for the task, and having facilitators circulate among the groups to ask and answer questions.

## SUMMARY

The training component of a citizen science project can play a huge role in the overall perception and understanding that participants have of the project as a whole. Training provides an opportunity to communicate and model the goals of the project, set the tone for the importance of the project, and frame the crucial role of participants. Moreover, training is an opportunity to model and foster a sense of camaraderie, community, and shared purpose, as well as a way to build social connections and have fun. We have offered suggestions here for planning and implementing training that is engaging, interactive, and reflective, thereby facilitating participants' learning of the protocols, skills, and content required to participate in your citizen science project. These techniques should also facilitate meaningful interactions and mutual learning among trainers and participants such that everyone walks away energized and enriched from the experience.

# NINE

# Collecting High-Quality Data

ALYCIA W. CRALL, DAVID MELLOR,
STEVEN GRAY, and GREGORY NEWMAN

Data collection lays the foundation of any scientific research study. High-quality data are essential to ensure that you can adequately address your research question and, more importantly, that you can confidently use these data to inform management and policy decisions. The scientific method demands a level of rigor that necessitates a lot of planning to ensure that the process selected for data collection is credible. In fact, scientists probably spend 40 percent of their time on any research project making decisions about how to design a study (i.e., the experimental design)[1] and determining which protocols to use to collect the data needed to address a particular research question. As a citizen science practitioner, proper planning will allow you to determine your project's goals and define the steps needed to get you there.

To help demonstrate the point of planning for data collection, let us look at a couple of key scenarios. Say that your local nature center wants to compare the plant diversity of three habitat types—forest, grassland, and wetland—in a neighboring natural area. You may decide to use a vegetation plot within each habitat and compare a diversity index from each one. For example, Shannon's diversity index is a commonly used metric that takes into account how many species are in an area and the abundance of each species. This comparison will likely involve performing a simple statistical test, such as an analysis of variance (ANOVA) and post hoc tests, across the diversity index for each habitat type. An ANOVA assumes random placement of your vegetation plots within each habitat to be valid. But what

---

1. Note that the focus of this chapter is on data collection, not experimental design. Though much of our discussion contains important aspects of experimental design, a full description of it is beyond the scope of this chapter.

if you do not consider these assumptions and how these data will need to be analyzed to address your specific research question? Your project participants might go out and spend hours, over many days or weeks, collecting data opportunistically by placing plots in areas that are easily accessible to the hiking trail that runs through the boundaries of the natural area. Only later, when the data are being compiled, do you realize that you have put numerous resources into a data-collection effort that has provided your project with unreliable data that do not help answer the original question you had in mind.

Let us consider another scenario. You are concerned that the level of monitoring required to ensure safe water quality in your community is no longer being conducted due to recent budget cuts within your state's regulatory agency. To address this concern, you organize a group of local volunteers to begin taking water-quality measurements using a protocol developed by a neighboring state. Your group measures common water-quality metrics to form an index (comprised of pH, temperature, and dissolved oxygen) at specific locations throughout your watershed. After a year of data collection, you send your data to the agency, only to discover that the agency is unwilling to accept your data for two reasons: you collected measurements inconsistent with their mandated protocols, and the agency is concerned about data quality. If you had communicated with the agency *prior to starting your data-collection activities*, you would likely have collected measurements consistent with their protocol and ensured that the data were structured in a manner consistent with the agency's current regulatory mandates.

Many citizen science projects find that if they do not plan properly to ensure adequate data quality, participants end up sending data to a black box and may lose motivation to participate. Who can blame them? Sadly, such scenarios in citizen science projects are common. In fact, a review of citizen science projects reported that very little volunteer-collected data ever makes it to a decision-making forum (Conrad and Hilchey 2011). Researchers who study citizen science and develop citizen science projects have begun to call attention to routine data-collection issues as a way to increase our knowledge of common pitfalls related to volunteer-generated data and establish common practices for dealing with reoccurring issues (Thornton and Leahy 2012).

So, to facilitate the planning process for your project, we suggest you consider the following questions:

What data are needed to answer your research and/or management questions?

Will you be conducting a controlled experiment or an observational study?

How will data be collected?

How will these data need to be analyzed to answer research and/or management questions?

If statistics are required, what are the assumptions of the statistical methods you plan to use?

Who will be using the data you collect? What are their data-quality requirements?

Are others already addressing the same or similar research and management questions using citizen science as a tool?

We will begin with the last question. If you answered yes, it will impact how you approach your other plans moving forward.

## EXISTING DATA-COLLECTION PROTOCOLS: DO NOT REINVENT THE WHEEL

Before developing a citizen science project, take some time to find out what projects are already out there that might have similar research goals. If similar projects exist, you may be able to adopt their methods or at least build on existing efforts. Citizen science is popular, and several directories for projects already exist (e.g., see CitizenScience.org and SciStarter. org for updated lists of projects by subject and audience). In some cases, the protocols used by these existing projects have been field tested with volunteers to ensure data quality and have been published in peer-reviewed journals (e.g., Genet and Sargent 2003, Delaney et al. 2008). Scientists working with local, state, or federal agencies may also be using data-collection procedures that you should consider for use in your project. A simple Google Scholar search for "citizen science" or "volunteer monitoring" and the subject you will be investigating will yield a list of potentially relevant reports or publications. However, as a citizen science practitioner, you may not have access to these publications without being charged a fee, in which case we recommend contacting the author directly for a copy of the publication. In most cases, authors are happy to provide one. Please keep in mind that any protocol you adopt, even if it has been used by another program, should still be assessed for the quality of data it produces to ensure your project's success.

Building on the existing efforts of others has multiple benefits, but from a scientific and data-quality perspective, one of the most important benefits is *standardization*. Standardization within and across citizen science projects allows scientists to collect data across large geographic areas and over long periods by merging data across individuals or groups. For these projects, everyone participating follows a standardized protocol, and the data collected are typically entered in an online database for data synergy and analysis (e.g., eBird, USA National Phenology Network). Even smaller local or regional projects need to consider the standardization of approaches for data comparison across years to analyze trends. These projects can also have significant impact by allowing data from local community efforts to be merged using standardized measures so that data collected across communities can be compared. For example, CitSci.org is an online database that allows users to effectively share measurement tools and observations across projects and learn from one another. On this website, project coordinators choose from an existing list of standardized measurements or, as needed, create their own measurements relevant to local goals and objectives. In this way,

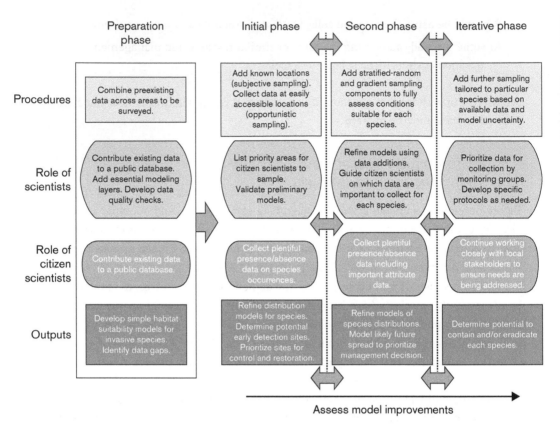

|  | Preparation phase | Initial phase | Second phase | Iterative phase |
|---|---|---|---|---|
| Procedures | Combine preexisting data across areas to be surveyed. | Add known locations (subjective sampling). Collect data at easily accessible locations (opportunistic sampling). | Add stratified-random and gradient sampling components to fully assess conditions suitable for each species. | Add further sampling tailored to particular species based on available data and model uncertainty. |
| Role of scientists | Contribute existing data to a public database. Add essential modeling layers. Develop data quality checks. | List priority areas for citizen scientists to sample. Validate preliminary models. | Refine models using data additions. Guide citizen scientists on which data are important to collect for each species. | Prioritize data for collection by monitoring groups. Develop specific protocols as needed. |
| Role of citizen scientists | Contribute existing data to a public database. | Collect plentiful presence/absence data on species occurrences. | Collect plentiful presence/absence data including important attribute data. | Continue working closely with local stakeholders to ensure needs are being addressed. |
| Outputs | Develop simple habitat suitability models for invasive species. Identify data gaps. | Refine distribution models for species. Determine potential early detection sites. Prioritize sites for control and restoration. | Refine models of species distributions. Model likely future spread to prioritize management decision. | Determine potential to contain and/or eradicate each species. |

Assess model improvements

FIGURE 9.1. Example of a data-collection protocol for invasive plant species, combining efforts of scientists and citizen scientists. Credit: figure reprinted, with permission, from Crall et al. (2010).

users share measurements and associated protocols across projects and thus standardize data-collection protocols, thereby allowing integration of data for across-project analyses.

Another important issue to consider as you develop a data-collection protocol is the degree to which you might like the data to be integrated into professional monitoring activities that are already in place. By *professional,* we refer to monitoring activities that are being conducted by paid scientists or natural resource managers as part of their work. There are trade-offs when collecting data with either volunteers or professionals, so integrating the two approaches can yield data more powerful than either would independently. For example, figure 9.1 shows the roles of scientists and citizen scientists in a collaborative invasive-species monitoring program (Crall et al. 2010). In the preparation phase, existing data (species location and habitat information) are combined across groups. Scientists can then use these data to get a better understanding of a species' known geographic distribution and produce habitat suitability models to determine areas where a species is likely to spread. This geographic information can then be provided to volunteers to help them target monitoring efforts at specific locations for collecting additional data on the presence or absence of a species (initial phase; figure 9.1). These new presence/absence data can then support further refinement of the habitat suitability models to inform more rigorous sampling by volunteers and professionals (second phase; figure 9.1). As more data are collected, iterative adjustments can be made to ensure that the protocols continue to produce data that meet the needs of each stakeholder.

As you develop any protocol, you should prioritize the steps necessary to initiate the proposed project within current resource constraints while outlining potential approaches for expansion as new resources become available. You should also seek to leverage resources available from local partners currently engaged in efforts similar to your proposed project. These resources include monitoring tools, online data-management systems, and established volunteer networks.

## MADE FROM SCRATCH: CREATING A NEW DATA-COLLECTION PLAN

Despite the advantages to your project and to the greater science community of using existing protocols, there may be compelling reasons to create your own—in some cases, appropriate protocols may simply not exist. Perhaps the species of interest is especially rare or behaves in ways that make existing protocols inappropriate, or the habitat in which you are conducting your investigation does not lend itself to existing methods. Perhaps the only protocols relevant to your study are suited to professional practitioners and cannot be easily translated into a citizen science framework. You could also be in uncharted territory and asking a question that no one, professional or volunteer, has asked before. Whatever the reason, creating your own protocol may be the best way to move forward as you consider your research questions.

As you begin, it will be necessary to follow best design practices with a focus on creating a high-quality dataset that could potentially withstand scientific review. Rushing through this design process has relegated datasets generated by many projects to the dustbin and, despite potentially providing a useful public outreach campaign and a good learning experience for volunteers, can lead to wasting time, money, and reputation among your fellow practitioners. Again, we recommend planning ahead to avoid these common pitfalls.

One big decision you will need to make before selecting a protocol is whether you want to conduct an *experimental* or an *observational* study. An experimental study is one in which you select a control (or untreated group) and then apply various treatments to your experimental groups. You then observe the effects of those treatments on each group in comparison to the control. For example, you may be interested in how increasing nitrogen in the soil might impact biomass in a native grassland. In this case you can set up a control vegetation plot of a certain size with no nitrogen applied and several treatment plots with four different levels of nitrogen applied. You can then measure biomass in each plot to determine which level of nitrogen resulted in more or less biomass.

How you design your experiment is extremely important, and care should be taken. Several things must be considered, including control, randomization, and replication. Establishing a control will limit experimental bias, which would favor certain outcomes over others. Thus, returning to our example, the control plot and treatment plots have other variables besides nitrogen that may impact biomass. These include rain, the amount of phosphorus in the soil, and the amount of sunlight the grassland receives. As a result, you not only want a control plot where you do not add nitrogen, but you also want these other variables to be as constant as possible across all your plots. Otherwise, you might determine that a certain application of nitrogen produced the most biomass when, in actuality, it was a combination of nitrogen and the amount of rain received in one of the treatments.

You may also determine that there needs to be randomization and replication in your experimental design. In our example, you have five plots within a homogeneous area of grassland, four of which will be supplied a treatment of nitrogen. You can randomly select the location of your plots within this area and also randomly assign their treatment. Replication is typically required as well to make the results of your experiment significant. Seeing similar differences across five plots replicated ten times instead of just once demonstrates greater validity in your findings. We recommend, as part of the experimental design process, consultation with a statistician to determine the number of replications needed to ensure an adequate sample for analysis.

In contrast to an experimental study, an observational study is one in which you make observations and measure variables of interest without performing any manipulations on treatments. Using our same example, you would randomly place plots within your grassland. Over time, you might collect data on the amounts of nitrogen deposition, phosphorus deposition, rainfall, sunlight, and biomass that occur within each plot. With these data, you cannot infer cause and effect, but you can make comparisons to see if common trends develop to determine what variable results in greater biomass.

## Protocols to Consider

Numerous resources are available for developing appropriate protocols to meet specific objectives. In fact, classes that pertain to measurement methods, sampling, and data analyses are a major focus of all scientific graduate training. As a result, many of the existing methodological resources have been written for professional researchers and may go beyond what you are trying to achieve through a citizen science project. Nevertheless, these can be modified or translated into a time- and labor-sensitive approach more suitable for a volunteer data collector. There are many approaches to consider when getting started. We highlight a few here, organized into levels, where each level increases the intensity—that is, the effort, time, and expertise required of volunteers. Likewise, the complexity of research questions that can be answered increases with the level of intensity. When deciding how intense an effort your study should have, be aware of the trade-offs inherent in this decision: less intensive protocols tend to require less training and are conducive to large-scale projects, or at least ones in which the ratio of trainers to trainees is low. However, the inverse is also true with highly intensive protocols.

Levels allow you to consider the broad range of knowledge and skills that different participants may bring to their volunteer service, while also allowing you to better integrate your data into professional efforts (as described in figure 9.1). Here, we will describe three levels for you to consider when developing your citizen science project. Level I includes ad hoc monitoring of a region and reporting of new data points. Level II requires more consistent measurements and rigorous methods, so that an area can be systematically evaluated. Level III includes experimental studies, in which researchers can determine the cause of a phenomenon. Although there is overlap between levels, the core questions that can be addressed with each one are basically as follows: (I) Is something present in this region? (II) Is there a pattern or a correlation between two or more measured variables in an area? and (III) Does a change in one variable *cause* a change in another variable?

These three levels of volunteer projects are directly linked to the ways that professional scientists think about evidence. Sometimes it is good enough to know whether something exists somewhere. Is a new invasive plant species coming into a national park? Can we have an early warning of the presence of higher-than-normal nutrient levels in a watershed? Other times, a researcher wants to know if two variables are related, so that the presence of one can predict the presence (or absence) of another. If invasive species are more likely to be found near a hiking trail, then systematic and comprehensive measurements are required near and far from the trails. If water quality is more likely to be low around construction sites, then regular measurements need to be made during all kinds of weather events.

At Level I, it is critical to collect the minimal amount of information that can be useful. The observation should include Who, What, Where, and When. Record information on who collected the data (the volunteer's name), what species are being recorded, where the specimen was found, and when it was recorded (figure 9.2). Level I monitoring should only focus on several target species or measurements to limit the amount of training needed and increase the number of observations made for those species or environmental attributes being evaluated.

We will go into more detail about data precision in a later section, but it bears mentioning here that each of the data points collected in response to the four questions (Who, What, Where, and When) could be recorded with varying levels of precision. For example, is it important to know the time of day that a species was observed or that a sample was taken, or will just the date suffice? For location, is just a general description of the area needed, or are GPS coordinates required? The answers to these questions will be evident if you have a clear understanding of your goals, based on your research question. Will somebody be using the data to field-check the observations or return for more thorough data collection? If so, precise location but imprecise species information may be okay. If observations will be used to create a regional model of species changes, then perhaps the precise location is not so important but species identification and abundance are.

A final comment about the sampling intensity of Level I: although the Level I design is limited in the types of questions that can be answered with it, this level of sampling becomes more powerful with larger sample sizes (i.e., more data-collection points). In fact, with the rise in use of smartphone apps for data collection, it is possible to amass a formidable

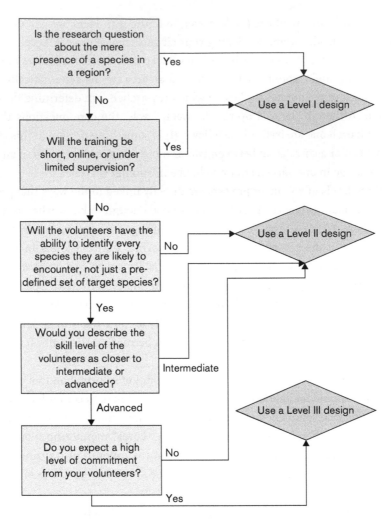

FIGURE 9.2. Determining the level of intensity for a citizen science data-collection protocol will depend both on the types of questions you are answering and on the skills, commitment, and experience of those working on your project. Level I monitoring will require the least training and experience, though fewer questions can be answered with data collected in this manner. Level III, on the other hand, requires extensive training and commitment from your volunteers, but opens up the range of possible research questions. Level II is a happy medium that many successful projects utilize, often using professionals to augment the data, with fewer Level III data collections.

dataset by asking a little from a lot of people. The intensity of effort at Level I is appropriate to answer questions about the presence of a species, pollutant, or activity in an area—in other words, "Does this exist here?" This question of existence is especially relevant for scientists and managers who want to find out where an emerging invasive species or a rare native species exists. However, it is inappropriate to use an opportunistic Level I sampling method if you want to ask questions about the rate of spread or the abundance of a species. For these types of questions a more structured effort is required.

At Level II, more data should be collected than the minimum as laid out for Level I, but the protocol will vary vastly depending on the nature of your project and research questions (figure 9.2). For instance, your focus may be on plants (or other stationary critters), animals, water testing, or habitats, and this will, of course, dictate the precise protocol design. Plant studies will often require setting up plots and taking data within them. On the other hand, animal surveys are likely to require taking data along transects, or lines, within the animal's expected range. Even if you are creating your own protocol, we still strongly recommend borrowing liberally from previous studies, or at least being familiar enough with those in your field to do so. There are several books that can help guide you through this process (e.g., Lindenmayer and Likens 2010, McComb et al. 2010).

Setting up research plots along transects requires that you determine the distance between transects, the distance along each transect that plots will be placed or samples taken, the number of subplots within each plot, the shape of the plot, and, in some cases, the time that will be spent at each plot (e.g., if you are conducting auditory surveys by listening for an animal's call for a predetermined amount of time). The answers to all these questions need to take into account both the scientific and volunteer requirements: How dense is the focus population or habitat, how many sampling points will collect the necessary information, how much effort can you request that each volunteer make, and how hard is it to get to locations and get through the habitat?

Let us consider a specific example of Levels I and II using invasive plant species. For this example, your organization is concerned about the presence of several invasive species that may occur on parkland recently acquired by the city. These species—common buckthorn (*Rhamnus cathartica*), Asian bittersweet (*Celastrus orbiculatus*), garlic mustard (*Alliaria petiolata*), and dame's rocket (*Hesperis matronalis*)—have been problematic in a neighboring park and volunteers have spent several years removing them. So, to begin, you just want to determine whether these four species are present in the new park. Using Level I sampling, the presence or absence of each of these species is recorded in an opportunistic fashion. In other words, volunteers record observations when one of these species is encountered as they walk through the park (figure 9.3). By collecting the data on your four target species, you can now answer research questions such as (1) Which of these species are currently present in the park? (2) How widespread are these species? and (3) What habitats are these species invading?

In the Level II protocol, you add the collection of species cover within three square subplots ($1 \text{ m}^2$) nested in a larger circular plot ($100 \text{ m}^2$). Within the circle, you record presence

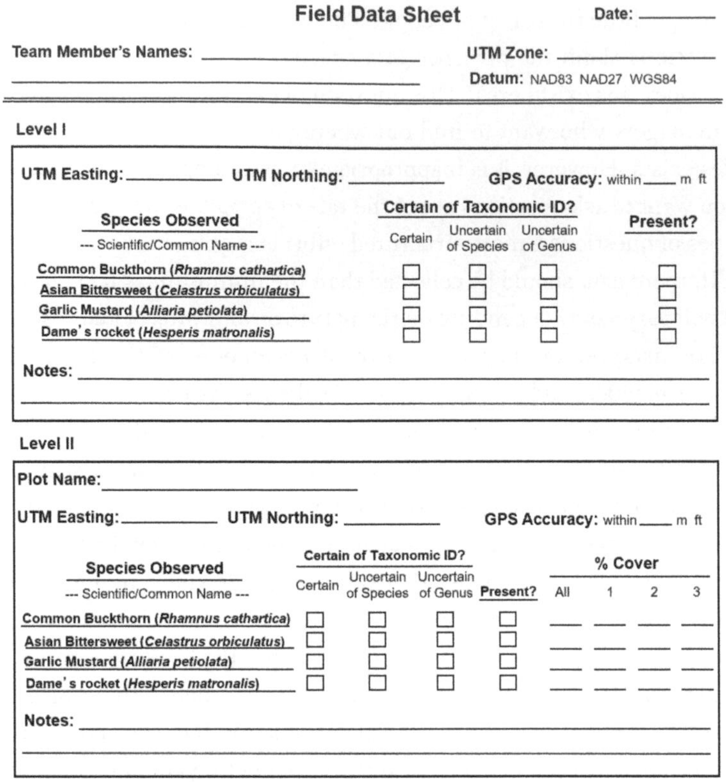

**Field Data Sheet**  Date: _____

Team Member's Names: _____  UTM Zone: _____
_____  **Datum:** NAD83 NAD27 WGS84

**Level I**

UTM Easting: _____ UTM Northing: _____  **GPS Accuracy:** within _____ m ft

| **Species Observed** | **Certain of Taxonomic ID?** | | | **Present?** |
|---|---|---|---|---|
| --- Scientific/Common Name --- | Certain | Uncertain of Species | Uncertain of Genus | |
| Common Buckthorn (*Rhamnus cathartica*) | ☐ | ☐ | ☐ | ☐ |
| Asian Bittersweet (*Celastrus orbiculatus*) | ☐ | ☐ | ☐ | ☐ |
| Garlic Mustard (*Alliaria petiolata*) | ☐ | ☐ | ☐ | ☐ |
| Dame's rocket (*Hesperis matronalis*) | ☐ | ☐ | ☐ | ☐ |

Notes: _____
_____

**Level II**

Plot Name: _____

UTM Easting: _____ UTM Northing: _____  **GPS Accuracy:** within _____ m ft

| **Species Observed** | **Certain of Taxonomic ID?** | | | | **% Cover** | | | |
|---|---|---|---|---|---|---|---|---|
| --- Scientific/Common Name --- | Certain | Uncertain of Species | Uncertain of Genus | **Present?** | All | 1 | 2 | 3 |
| Common Buckthorn (*Rhamnus cathartica*) | ☐ | ☐ | ☐ | ☐ | ___ | ___ | ___ | ___ |
| Asian Bittersweet (*Celastrus orbiculatus*) | ☐ | ☐ | ☐ | ☐ | ___ | ___ | ___ | ___ |
| Garlic Mustard (*Alliaria petiolata*) | ☐ | ☐ | ☐ | ☐ | ___ | ___ | ___ | ___ |
| Dame's rocket (*Hesperis matronalis*) | ☐ | ☐ | ☐ | ☐ | ___ | ___ | ___ | ___ |

Notes: _____
_____

FIGURE 9.3. Example data-collection sheet for Level I and Level II monitoring. In this example, a volunteer will be trained to identify four invasive plant species of concern to local natural resource managers. The volunteer team records their names (Who) and the date (When) before starting a hike through the natural area. As any of the four species are encountered, they record it as present (What) and provide a location by reading UTM coordinates from GPS (Where). For Level II, plots are added to the protocol and cover estimates are recorded.

of the species. In the three subplots, you record cover estimates to measure species abundance. By recording the exact location of the plot and subplots, you can return to the same location year after year to look at trends over time (figure 9.3). With these data, you can now address the following research questions: (1) Is the population of these species increasing or decreasing over time? (2) Does the abundance of a species differ in different habitats?

A Level III protocol might take a specific hypothesis and put it to the test with an experimental design (figure 9.2). Suppose that you notice a lot of invasive plant species on trails where dogs are allowed. You think that dogs may inadvertently be bringing in seeds on their coats, but perhaps there is something else going on. Perhaps pets are allowed only in parks closer to cities, and that is why there are more invasive plant species where there are also more dogs. Or perhaps pets are allowed only in parks where there are no known rare or pristine ecosystems. In that case, it would be the opposite causal relationship: invasive plants allow dogs to be present. In order to be sure there is no unknown variable causing the increase in invasive plants, the scientist would have a much tougher time conducting the investigation. She would have to be able to randomly assign some parks to allow dogs and others to prohibit them. The challenge with these experiments is, of course, the fact that few scientists (professional or nonprofessional) have the power to make huge decisions that could address the question at hand. Some questions are answerable, whereas others have to rely on the best evidence that a Level II protocol can provide.

### SAMPLING DESIGN AND HOW TO AVOID SAMPLING BIAS

Regardless of which protocol you use, it will require putting some thought into sampling design. Specifically, there are three major sampling designs to consider, each with its own traits: *random, systematic,* and *stratified random.* In random sampling, you use a tool, such as a computer-generated random number table or even a phone book, to randomly pick sampling locations within a defined area (figure 9.4A). In systematic sampling, you sample at a consistent interval throughout the range (figure 9.4B). Notably, with both random and systematic designs, it is possible to under-sample or not sample a habitat of interest. Stratified random sampling ensures that you randomly sample within a given habitat type, but you guarantee a minimum number of visits in each habitat of interest (figure 9.4C). Whatever your study requires, the same basic principle will guide any investigation and must underlie the design you choose.

Why is random sampling so important? Because it is often impossible to collect the data that would characterize an entire population,[2] statisticians have devised ways to make reasonable inferences from a small subset of data. Although having all of the information about the complete population would allow us to characterize a phenomenon completely, it is usually cost-prohibitive, if not impossible, to sample every blade of grass, or find every rare salamander, or map every single pollinator-rich habitat. In fact, in the case of species, we know the population size of only a handful of them, such as those living on small islands or in captivity. Given that we must use less than complete knowledge to gain insight about the whole system, how do we go about collecting that information? The key is to avoid *bias*.

When you sample a subset from a population, you risk inadvertently capturing individuals that are slightly different than the true average of normal individuals, in which case your data would be *inaccurate*. We will discuss collecting high-quality data in more detail below (and see chapter 9), but the following analogy is important for discussing bias in data sampling. Imagine looking over a field for a particular species of rare flowering plant. Without much forethought, it would be easy to just find the tallest or most colorful individuals. Then,

2. By *population*, we mean 100 percent of a group or item of interest, not simply the total number of individuals of a species. That is, the term *population* is used here in a statistical context, meaning the complete sample or total of an item.

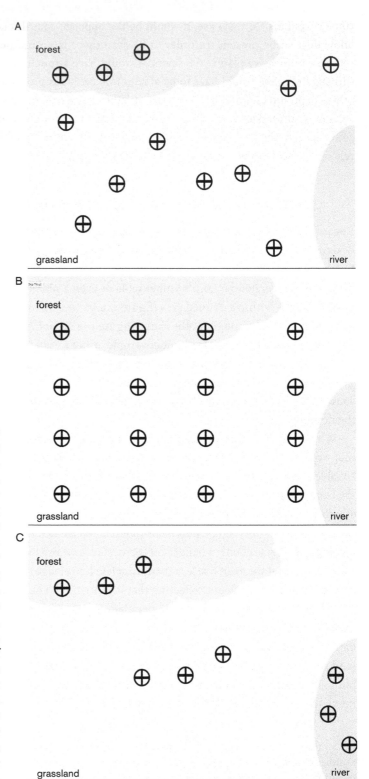

FIGURE 9.4. Three commonly used sampling designs. Although both random sampling (**A**) and systematic sampling (**B**) are widely used to collect field data, both may miss small habitat types within your sampling area. By randomly sampling within every known habitat type, as is done with stratified random sampling (**C**), you maintain the benefits of random sampling while ensuring that each habitat is represented in the final dataset. This assumes that you have prior knowledge of the habitat types when planning your data-collection process.

if you were to make inferences about the whole plant population from that biased subset, your inferences would likely be skewed. Likewise, imagine conducting a species survey along a trail or roadside, because access to the interior of the habitat is troublesome. The species that exist near the edge of the habitat are likely to be different from those that live in the interior, so your sample will not reflect the actual population residing in the region.

Every study will likely have some bias associated with it. This is almost unavoidable. However, minimizing bias is always crucial, because collecting an unrepresentative sample leads to inaccurate knowledge—simply put, it is bad science. Make every effort to remove bias at this planning step, because seemingly trivial differences often lead to unanticipated outcomes. Lastly, when the limitations of your study design require some bias, as they inevitably will, be certain to clearly state these biases and limitations up front. It may seem like you are admitting a defeat, but in reality you are allowing others to make a more accurate inference from your work. Hopefully, future investigators will be able to remove such biases with new, creative sampling designs or by using post hoc statistical approaches.

If biases are known and quantified, then it may be possible to address them using a statistical approach. For example, an analysis of covariance (ANCOVA) is similar to an ANOVA, though it uses linear regression to account for a confounding variable, the covariate. However, in order to use the ANCOVA, you must have identified and measured the variable that may be introducing bias. Notably, there are other assumptions that the ANCOVA test makes, and it is beyond the scope of this chapter to provide a comprehensive list of these assumptions—or of other statistical tests, their assumptions, and how to conduct them. Thus, we recommend purchasing a book on statistics or consulting a statistician early on in your design process so that you can be certain that your design allows you to answer your research questions. In the life sciences, *Biometry* by Robert R. Sokal and F. James Rohlf is the gold standard for using statistics, though it is a bit dense for a beginner (Sokal and Rohlf 1995). A very user-friendly book that helps practitioners choose a statistical test and use it in common software packages is Calvin Dytham's *Choosing and Using Statistics: A Biologist's Guide* (Dytham 1999).

### CHECK ALL THAT APPLY: METHODS FOR DATA COLLECTION AND ENTRY

Now that you have devised a sampling design that minimizes bias, you need to begin collecting your data. Data collection includes making observations in the field or lab but also recording and making those data available to your project team and other interested stakeholders. In addition, data collection requires that volunteers have the necessary tools and training (see chapter 8) to do the job, not only in terms of accurately measuring variables, but also for entering observations. The scale of your project will introduce different sets of challenges for data collection. For example, small-scale projects may benefit from better collaboration through face-to-face interactions, but large-scale projects may have more resources and expertise to develop sophisticated online and smartphone data-collection tools. Obviously, using tools that most people have around their house is great, but be careful and mindful of choosing tools that will yield the quality of data your project requires. In addition, try to be aware of those you may be excluding when you require high-speed Internet connections, smartphones, or even large amounts of free time for participating. Although these resources may be necessary and your goals may not include reaching diverse participants, broadening participation in science is a frequently cited, but rarely achieved, benefit of citizen science. As practitioners, we should try to be as inclusive as possible.

If special equipment is required that may be beyond what the average person has at home, consider providing resources to your participants. For example, the Community Collaborative Rain, Hail, and Snow Network (CoCoRaHS; www.cocorahs.org) requires that participants buy a high-capacity (four-inch-diameter) rain gauge to collect precipitation data. These rain gauges can be ordered from the project site and mailed directly to the participant's home. Likewise, the development of resource libraries is an approach taken in Wisconsin to provide monitoring resources for local groups. The libraries include GPS units, vegetation plots, and field guides focused on species of interest and are located in areas where volunteers are actively engaged in monitoring efforts.

There are numerous options available for recording data. However, as with every decision related to creating a protocol, each approach has its strengths and weaknesses. Identifying these strengths and weaknesses, and your audience, early on will help you decide how to proceed. Data-entry tools essentially exist on the same spectrum as data-collection protocols: from completely prepacked (i.e., use what others have created before) to fully customized (i.e., make your own). In between are a suite of tools and platforms, including those designed specifically for the needs of citizen science practitioners. Below are brief descriptions of several data-entry tools and their associated costs and benefits.

Paper-and-pencil data-entry forms should not be overlooked in the zeal to use fun and sexy mobile apps. There are many advantages to sending out your volunteers armed with data sheets and having them mail or email them back to you. Hard copies never run out of batteries, are easy to create and understand, and for many it is reassuring to have a physical copy of the raw data always available. However, data sheets may easily get lost, poor handwriting can make some data essentially useless, forms may never be mailed in, and transposing data into a digital format can be a costly burden for both small and large projects with limited staff and resources.

Web platforms that allow volunteers to enter data on their own solve many of the shortcomings of using paper-only data sheets but come with their own costs. Generic, cloud-based data sharing and storage tools such as Google Docs and Evernote are a simple way to share and receive information from volunteers. Custom-built websites can be costly to create and manage, but they can be optimized to have specific features your project may require. There are even customizable platforms designed specifically for citizen science projects, such as CitSci.org and National Geographic's FieldScope. However, having volunteers enter their data into a computer after they collected it in the field assumes that volunteers will always be diligent in doing so, when in fact some may not follow these instructions.

Smartphone apps allow volunteers to submit data directly from the field, potentially minimizing the loss of data. However, these apps require tech-savvy volunteers who own a device compatible with whatever app you decide to use. Furthermore, participants must be willing to take a device into the field where data are to be collected. While in the field, mobile devices can be difficult to read and may not have the reception required to receive or transmit the necessary information. As with web platforms, both general and customizable mobile citizen science platforms exist. Be cognizant that many mobile apps are limited to either Apple iOS or Google's Android operating system. You can create your own smartphone apps to collect the data you need in a way specific to your project's needs, but the cost and expertise required to develop your own apps tend to be expensive and specialized, much like developing your own website.

### ENSURING HIGH-QUALITY DATA: KNOW YOUR AUDIENCE

Data quality also needs to be considered in the early stages of protocol development for data collection. Specifically, data collected through your project may be used to guide research, management, or policy related to what is being studied. Concern over data quality in citizen science projects has sparked a significant number of peer-reviewed publications comparing data collected by citizen scientists to data collected by professionals (e.g., Foster-Smith and Evans 2003, Lewandowski and Specht 2015). These publications span multiple fields and multiple protocols, but the take-home message for most is that citizen science can provide high-quality data when appropriate protocols and data-quality procedures are in place. Please note that skepticism about the quality of the data that result from citizen science projects continues to run high (Nature Editorial 2015). Thus, it is imperative that you take the necessary steps to ensure data reliability. As a growing community, practitioners of citizen science should feel a sense of responsibility to others in the field in order to enhance the reputation of the work that comes from it.

In discussing data quality, there are two key terms that warrant consideration: *accuracy* and *precision*. An easy way to describe and differentiate accuracy and precision is by using the following analogy. Let us say you are trying to hit the bull's-eye on a dartboard. Your first throw of the dart lands in the top center of the board, missing your target. However, your next three throws all land next to your first throw. Although each throw was inaccurate (i.e., did not hit the intended target, the bull's-eye), the throws were precise (i.e., repeated throws produced a similar result; figure 9.5). In other words, accuracy is how close the dart (or, in data-collection parlance, the sample) is to the bull's-eye (or the true or correct value), while precision is how consistent or repeatable the dart throws are in relation to one another (or how consistently you collect or measure a sample). When collecting data for scientific research, you want to strive for both accuracy and precision through your data-collection protocol.

What specific steps can you take to ensure that the data you generate through your citizen science project are accurate and precise? The first approach is a strategy we have

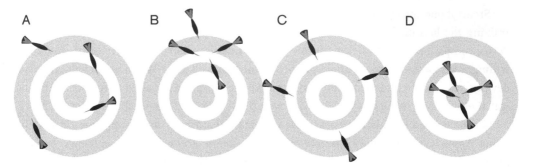

FIGURE 9.5. Dartboards showing examples of precision and accuracy. (A) The darts are randomly scattered around the board and are thus imprecise and inaccurate. (B) All the darts hit the upper center portion of the board, exhibiting precision but not accuracy. (C) The darts are spaced equally apart and are an equal distance from the target, demonstrating accuracy because—mathematically, the average of the darts would fall within the bull's-eye (the target). (D) The dart thrower was both accurate (hitting the bull's-eye) and precise (repeated throws produced similar results).

advocated before: to build on the efforts of others. If you are addressing a research question that can use an existing protocol that has been shown to be effective using volunteers, use it. Using existing protocols will save you a lot of time and help standardize data collection, as discussed earlier in this chapter.

Proper training is also an effective approach to ensure data quality and is discussed at length in chapter 8. Specific skills will be required to implement a protocol, and you need to be sure that training provides participants with the level of skill required to produce high-quality data. Skills might include georeferencing locations, species identification, and use of specialized equipment. Do not assume that volunteers will gain the necessary skills needed from any form of training. Evaluation is a valuable tool for determining whether participants complete a training with the knowledge, skills, and confidence to participate effectively in your program (chapter 12). Such evaluation demonstrates your due diligence in making the necessary modifications to your training and protocol to achieve your goals.

Many volunteer programs have started using online data-management systems to collect, store, and disseminate data from large-scale citizen science programs. Some of these systems have internal methods for ensuring data quality. For example, eBird programmers have developed smart filters to flag suspect data contributed online (Bonter and Cooper 2012). New observations are flagged if they fall outside of the known distribution and/or abundance for a specific species at a given time of year. Another online data-management system, designed through the Great Lakes Early Detection Network, sends new observations of invasive plant species submitted by volunteer observers to a group of verifiers that are registered to receive sightings in a defined area. Once verified, the data are made available to anyone accessing the website (Crall et al. 2012).

In addition, data-management systems like iNaturalist use crowdsourcing approaches to increase confidence in species identifications. For any observation, multiple participants can verify that the sighting reported was accurate. This multi-observer verification technique can be replicated in smaller projects by having about three independent observations for any given point. Any aberrant observation can be flagged and you can determine its validity on a case-by-case basis or by following a predetermined set of rules for exclusion. This multi-observer verification method highlights the trade-offs necessary to ensure high-quality data: more repeated observations require fewer total observation points, and you will have to decide how to balance the conflicting priorities. Additional methods might include drop-down pick lists that eliminate typos or the use of smartphone apps that automatically submit location coordinates. New technologies will continue to provide resources to improve data quality, so we encourage readers to research the latest software and tools available for this purpose.

## SUMMARY

Not everyone who contributes to scientific research needs to be a professionally trained scientist, but using the scientific method as a tool for understanding our world does come with some expectations and standards. These standards and expectations allow us to build knowledge continuously as new discoveries about our world are integrated into a foundation of discoveries that came before. Scientific research conducted in partnership with volunteers should continue this history of discovery.

In this chapter, we have presented considerations and guidelines to help ensure that the discoveries made by citizen science will contribute to our understanding of the natural world and be used in meaningful ways. Although it is difficult to anticipate the many issues associated with working with volunteers in scientific research, there is a growing literature on common pitfalls, specifically with regard to data quality, that need to be considered in project design. However, learning from similar projects and building on existing and robust data-collection protocols while anticipating how the data generated are going to be collected, analyzed, and ultimately used to make inferences should help project designers avoid issues associated with data quality and ensure that citizen science will remain a valuable tool for the scientific process into the future.

# TEN

# Data Management and Visualization

GREGORY NEWMAN, SARAH NEWMAN, RUSSELL SCARPINO,
NICOLE KAPLAN, ALYCIA W. CRALL, and STACY LYNN

Managing data is akin to organizing and managing items in your closet. Without proper planning and continuous attention, the number of items grows over time and becomes unwieldy, lacking efficient labeling, organization, compartmentalization, preservation, and storage procedures. When you do not take time to manage the chaos, it gets increasingly more difficult to find and access items and identify which item is right for your needs at any given time. However, you can label bins with consistent naming conventions to improve your ability to find the most appropriate items, while taking care to preserve them (e.g., the addition of cedar chips to sweater bins). In fact, your planning might involve hanging a map on the inside of the door that visually depicts where to find various items and how many of each type exist, and you might use names to indicate how they are organized and related.

Effective data management facilitates easier understanding of information in the form of meaningful analyses and visualizations. After all, science, and therefore citizen science, is about advancing our understanding of the world by making discoveries and then sharing those discoveries with the community. The foundation for a good discovery lies in effective data management, analysis, and visualization that arise from well-designed data-collection protocols (chapter 9). In this chapter, we define and examine the concept of data, walk through the stages and planning of the data life cycle, discuss approaches and techniques of data visualization, and share guidelines for data management and visualization that we find helpful for citizen science practitioners and project managers.

## THE CONCEPT OF DATA

The concept of data is described and defined from several viewpoints. In the context of citizen science, we define data as information translated into a form that is easily moved, processed, or visualized (typically in binary digital formats) and that generally represents a gathered body of facts. The types of information gathered include diverse formats such as images, videos, numerical data (integer, decimal), and textual data (character strings). Thinking broadly, data also includes songs, music, pictures, videos, numbers, prose, and discourse and is a concept related to the notions of information, knowledge, and wisdom. For instance, the height of Mt. Everest and a photo taken of the summit are considered data, whereas a book on Mt. Everest's geological characteristics is considered information, and a report containing practical information on the best way to reach the peak might be considered knowledge. In citizen science, data collected by participants can take the form of any of these formats and often are representative of many formats within a single project. For example, an observation of a bat might include a photo, a sound recording obtained from an Anabat detector, textual comments made by a citizen scientist, GPS location data, the date of the observation, and a count of the number of bats observed. For another example, data presented to participants in a crowdsourcing citizen science project designed to interpret the shapes of various galaxies might include the data presented to the participants as images taken from a satellite or telescope, as well as categorical data representing the choices made by participants classifying the imagery. In either case, and across many more types of citizen science projects, effective management, analysis, and visualization of data can go a long way toward improving the overall success, impacts, and outcomes of your citizen science project (Wiggins et al. 2011).

## DATA MANAGEMENT

Data management is the process of controlling (and organizing) the information generated during your project, which includes the development, execution, and supervision of plans, policies, programs, and practices that control, protect, deliver, and enhance the value of data and information assets. Data life cycle management (DLM) is a useful, policy-based approach to managing the flow of an information system's data throughout its life cycle: from creation and initial storage to the time when the data become obsolete and are deleted. In citizen science, many DLM activities occur in parallel to the tasks and steps of the scientific method (figure 10.1). But before we delve into the data life cycle and its associated tasks, activities, and steps, we should ask: Why is it important to manage data, anyway? The answer is that effective and efficient data management

facilitates data collection, compilation, and sharing;

makes data more understandable and reusable;

avoids data loss;

facilitates easy analysis and visualization;

meets grant requirements;

enables replication of research;

ensures the legacy of your project long into the future;

facilitates serendipitous discoveries;

streamlines the identification of outliers and errors;

increases the rigor of scientific (including citizen science) research; and

saves time and money.

These are just a few of the many benefits of effective data management. In general, it is important to manage data because doing so will benefit the scientific community, including you and your collaborators, and will be important for publishing in scientific journals and meeting the requirements of funders or sponsors of the research (Strasser et al. 2011).

### THE DATA LIFE CYCLE

Central to the process of data management is the *data life cycle*—the stages of project data management that occur throughout a project, from start to finish and beyond (figure 10.1). This life cycle is important because it guides the decisions you make regarding data throughout your project. The data life cycle reminds you about important steps needed to organize information with the goals of analysis, visualization, sharing, and reuse in mind. To accomplish such organization, the life cycle consists of eight stages: Plan, Collect, Assure, Analyze, Describe, Preserve, Discover, and Integrate (figure 10.1; Michener and Jones 2011, Strasser et al. 2011, DataONE 2019a).

Stage 1 (Plan) involves writing a data management plan describing the dataset that will be generated and how it will be managed, documented, archived, and made accessible throughout its lifetime. Many federal granting agencies (e.g., the National Science Foundation) now require data management plans for research projects before they will award funding. The contents of the data management plan should include

types of data to be authored;

standards to be applied (e.g., format and metadata content);

provisions for archiving and preserving data;

access policies and provisions; and

plans for eventual transition or termination of data (if relevant).[1]

1. See www.dataone.org.

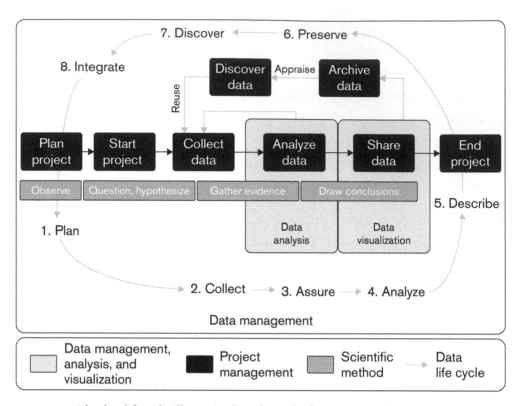

FIGURE 10.1. The data life cycle, illustrating how the work of managing, analyzing, and visualizing data is related to project management and the scientific method. Opportunities for appraisal, reuse, repurposing, and reappraisal of data are shown along with the important, yet often forgotten, project-management activities of data discovery and data archiving. Credit: adapted from University of Virginia Library (2019) and DataONE (2019a).

For example, you might jot down an inventory and estimated quantity of the data you expect to gather, the naming conventions you plan to follow, which software you envision using, and where you will store and back up your data.

Stages 2 and 3 (Collect and Assure) are activities you and your volunteers perform routinely: collecting data and ensuring data quality. These were discussed in detail in chapter 9, but it is worth reiterating that an experimental design and protocol well suited to your scientific research questions is an important first step toward generating and managing rigorous data.

Stages 4 and 5 (Describe and Preserve) ensure the long-term usefulness of your dataset so that it can be reused again in the future, either within or beyond the scope of your original intent. These fourth and fifth steps involve documenting data and changing data formats. For example, if your predecessors forgot to describe what was done and then stored their data on old 5¼ inch floppy disks, how would you make use of those historical data today? Data description and preservation help prepare for such realities and feed into the next two steps, data discovery and integration. Stage 6 (Discover) involves finding data that may be useful to your research. Stage 7 (Integrate) brings new information into your exist-

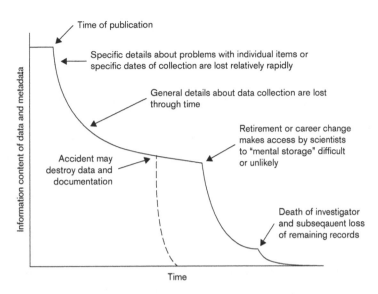

Time of publication

Specific details about problems with individual items or
specific dates of collection are lost relatively rapidly

General details about data collection are lost
through time

Retirement or career change
makes access by scientists
to "mental storage" difficult
or unlikely

Accident may
destroy data and
documentation

Death of investigator
and subseqauent loss
of remaining records

Information content of data and metadata

Time

FIGURE 10.2. The normal degradation of data, illustrating the demise
of data and metadata over time as a result of various circumstances.
Credit: reproduced, with permission, from Michener et al. (1997).

ing dataset. Finally, Stage 8 (Analyze) includes evaluating, interpreting, and assessing data
to yield useful information as results. We use analysis to evaluate whether our data support
the questions we sought to address.

By following this data life cycle, you position your data well for repurposing, reappraisal,
and reuse by yourself and others to make new discoveries and prevent the degradation of data
(figure 10.2). The key is to pay attention to each of these steps *at the beginning and continuously throughout* your citizen science project. Waiting until your project ends—a common
mistake—is too late. Once good data-management practices have become routine, many benefits suddenly emerge. In the following sections, we will delve into some of these benefits in
more detail, describe some techniques and tips, and provide several guidelines. But first, let's
take a look at the benefits of proper data management at each stage of the data life cycle.

### Plan

Managing the diversity of data types in a citizen science project (or any project) is fraught
with challenges and requires proactive planning. Computers increase efficiency and effectiveness but require consistent patterns and structure. As we learned in chapter 9, data are
ideally collected following carefully designed protocols geared toward specific research questions. If you want computers to be able to automatically manage, analyze, and visualize data,
your data must be consistent and well structured. The structure of your dataset is referred
to as its *schema*. The most common schema is a tabular structure (i.e., a table). Tables store
data in rows and columns, with the content of each cell defined by its column or row header.
Structures that are more sophisticated involve relational database-management systems

(e.g., Microsoft Access, MS SQL Server, and MySQL) that relate information stored in one table with information stored in another. Your choices of what information goes into which table, and into which row and column of a table, will define your data schema. When making decisions on data structure, it is important to consider whether you need to document provenance (e.g., changes made in your data through time or knowledge of the original source) and what format to use for measurements (e.g., integer, whole number, decimal, categorical, raster-based image, vector-based image, video, or text). For example, do you need to know that the initial observation made by José on September 3, 2016, reported a western bluebird *(Sialia mexicana),* but later Samantha changed this observation to Steller's jay *(Cyanocitta stelleri)* on the basis of José's photo? Should you record pH measurements as integers restricted to 1–14 in whole numbers or allow the reported values to be decimals?

Whether you use a simple spreadsheet, a more complex relational database, or a website, you must choose how to arrange your data. Chances are you will gravitate to a simple approach. For example, say you are seeking to compare the plant diversity of forests, grasslands, and wetlands. You randomly locate sixty plots (twenty in each habitat) and use the data you collect there to generate a species diversity index for each plot. A simple schema to organize these data could be the spreadsheet shown in figure 10.3A. In this spreadsheet, each row represents an observation of a plot and the corresponding species observed. Notice, however, that with this approach, you repeat information—the only data that change are the species observed. Were there really three separate soil pH measurements made at Plot 3 with a value of 5.1 (e.g., circled values in figure 10.3A)? Alternatively, you could organize these data in a flat file (figure 10.3B), in which each row represents a single unit of analysis (in this case, a single plot). Finally, when using a relational database (figure 10.3C), there is no need to repeat records because such data are stored only once in separate, related tables linked to plot data via a unique identifier. Regardless of the approach you have chosen, the next step is to begin thinking about collecting the data you will put into your schemas, spreadsheets, and databases.

### Collect: Tools, Approaches, and Software

Managing data well requires forming good habits and having the right tools, approaches, and software in your toolbox to get the job done correctly. The data you collect and manage may be quite diverse. For example, your data may include information about volunteers, the cost of equipment, actual field observations, pictures, videos, questionnaires, and even specimens. Given this variety of data, your tools may need to be equally diverse.

In general, tools consist of hardware, software, and materials to help organize information. Spreadsheets, geographic information systems (GIS), reference management software, online data-management systems, file cabinets, and even sticky notes all help manage data (table 10.1; see also DataONE 2019b). In addition to obvious tools, such as Microsoft Excel, there are websites that support data management for specific topics and protocols (e.g., Project Budburst for plant phenology, CoCoRaHS for precipitation, and eBird for bird observations) and still others that support the data-management needs of many citizen science projects regardless of topic (e.g., CitSci.org; see table 10.1 for more examples; Graham et al.

### (A) Spreadsheet – one row per "species"

| | | "Boiler Plate" Information | | | | | Organism Measurement | Site Characteristic |
|---|---|---|---|---|---|---|---|---|
| ID | Date | Habitat | Plot Name | Latitude | Longitude | Scientific Name | Percent Cover | Soil pH |
| 729 | 04/07/2014 | Wetland | Plot 1 | 41.529863 | -104.276483 | Juncus effusus | 14 | 4.5 |
| 729 | 04/07/2014 | Wetland | Plot 1 | 41.529863 | -104.276483 | Poa pratense | 15 | 4.5 |
| 729 | 04/07/2014 | Wetland | Plot 1 | 41.529863 | -104.276483 | Carex occidentalis | 61 | 4.5 |
| 729 | 04/07/2014 | Wetland | Plot 1 | 41.529863 | -104.276483 | Acer glabrum | 10 | 4.5 |
| 730 | 05/05/2014 | Forest | Plot 2 | 38.895432 | -98.889954 | Pinus ponderosa | 42 | 7.2 |
| 730 | 05/05/2014 | Forest | Plot 2 | 38.895432 | -98.889954 | Acer glabrum | 58 | 7.2 |
| 731 | 03/06/2014 | Wetland | Plot 3 | 40.189654 | -101.183492 | Juncus effusus | 67 | 5.1 |
| 731 | 03/06/2014 | Wetland | Plot 3 | 40.189654 | -101.183492 | Carex occidentalis | 23 | 5.1 (?) |
| 731 | 03/06/2014 | Wetland | Plot 3 | 40.189654 | -101.183492 | Poa pratense | 10 | 5.1 |
| 732 | 09/06/2014 | Grassland | Plot 4 | 37.929853 | -99.285491 | Poa pratense | 100 | 6.8 |

### (B) Spreadsheet – Flat File – one row per "observation"

| ID | Date | Habitat | Plot Name | Latitude | Longitude | Juncus effusus % cover | Poa pratense % cover | Carex occidentalis % cover | Pinus ponderosa % cover | Acer glabrum % cover | # of species | Soil pH |
|---|---|---|---|---|---|---|---|---|---|---|---|---|
| 729 | 04/07/2014 | Wetland | Plot 1 | 41.529863 | -104.276483 | 14 | 15 | 61 | | 10 | 4 | 4.5 |
| 730 | 05/05/2005 | Forest | Plot 2 | 38.895432 | -98.889954 | | | | 42 | 58 | 2 | 7.2 |
| 731 | 03/06/2005 | Wetland | Plot 3 | 40.189654 | -101.183492 | 67 | 10 | 23 | | | 3 | 5.1 |
| 732 | 6/9/2-14 | Grassland | Plot 4 | 37.929853 | -99.285491 | | 100 | | | | 1 | 6.8 |

### (C) Relational Database – many related tables

| ID | Date | Habitat | Name | Latitude | Longitude |
|---|---|---|---|---|---|
| 729 | 04/07/2014 | Wetland | Plot 1 | 41.529863 | -104.276483 |
| 730 | 05/05/2005 | Forest | Plot 2 | 38.895432 | -98.889954 |
| 731 | 03/06/2005 | Wetland | Plot 3 | 40.189654 | -101.183492 |
| 732 | 6/9/2-14 | Grassland | Plot 4 | 37.929853 | -99.285491 |

| ID | Scientific Name | PlotID | Percent Cover |
|---|---|---|---|
| 53 | Juncus effusus | 731 | 67 |
| 54 | Carex occidentalis | 731 | 23 |
| 55 | Poa pratense | 731 | 10 |

| ID | PlotID | Soli pH |
|---|---|---|
| 53 | 731 | 5.1 |

FIGURE 10.3. Three data structures for a forest/grassland/wetland diversity comparison study. (A) Raw data are stored in a spreadsheet by species and plot, replicating soil pH measurements for each plot while showing each species' unique percent cover data. (B) The same data, again in a spreadsheet, shown as a flat file; data for each plot are in a single row with a single soil pH value, species percent cover as individual columns for each species, and a derived response variable (e.g., the number of species found per plot). (C) The same information stored in a relational database consisting of three related tables. Notice that metadata are not explicitly contained within either structure and must be documented elsewhere, typically in associated metadata files.

TABLE 10.1   Tools for Data Management, Analysis, and Visualization

Tools are listed in alphabetical order and no preference by the authors is implied; those marked with an asterisk are free.

| Tool by category | Description | URL |
| --- | --- | --- |
| **Online systems** | | |
| AnecData.org* | A free online platform where anyone can create an account to start crowdsourcing georeferenced environmental and ecological data, supporting anything from water-quality data to full, multidimensional species biodiversity surveys | www.anecdata.org |
| CitSci.org* | A comprehensive online support system for citizen science projects that helps you create your own projects; create custom data sheets, analyses, maps, and visualizations; and use associated synchronized mobile apps | www.citsci.org |
| Indicia* | A toolkit that simplifies the construction of new websites that allow data entry, mapping, and reporting of wildlife records | www.indicia.org.uk |
| Ohmage | An open-source, participatory sensing technology platform that supports campaign authoring, mobile-phone-based data capture through inquiry-based surveys, and automated data capture | http://ohmage.org |
| Sensr | A web-based visual environment where people who want to collect and explore small sets of data can build an iPhone application as a data-collection tool | www.sensr.org |
| **Topic-specific systems** | | |
| ClimateWatch (phenology for Australia)* | A project to understand how changes in temperature and rainfall are affecting the seasonal behavior of Australia's plants and animals | www.climatewatch.org.au |
| iNaturalist (species observations)* | A community for reporting personal observations of any plant or animal species in the world | www.inaturalist.org |
| Nature's Notebook (plant and animal phenology)* | A project that gathers information on plant and animal phenology across the United States to be used for decision making on local, national, and global scales | www.usanpn.org /natures_notebook |
| Project Budburst (plant phenology)* | A network of people across the United States who monitor plants as seasons change; data are reported on the Project Budburst website and made freely available for everyone | www.budburst.org |

| | | |
|---|---|---|
| Zooniverse platform (topics vary)* | A citizen science web portal that hosts large-scale projects that allow volunteers to participate in online, crowdsourced scientific research | www.zooniverse.org |

**Spreadsheets**

| | | |
|---|---|---|
| Google Sheets* | A free application to create new spreadsheets and open existing ones that can be shared with and edited by others | www.google.com/sheets /about |
| Microsoft Excel | Software that allows users to organize, format, and calculate data with formulas, using a spreadsheet system broken up by rows and columns | https://office.microsoft. com/en-us/excel |
| OpenOffice* | An open-source office-productivity software suite | www.openoffice.org |

**Databases**

| | | |
|---|---|---|
| MySQL* | An open-source, relational database-management system | www.mysql.com |
| Microsoft Access | A database-management system that combines the relational Microsoft Jet Database Engine with a graphical user interface | www.microsoft.com /en-us/cloud/products /sql-server-editions /overview.aspx |
| Microsoft SQL Server | An enterprise-level, relational database-management system whose primary function is to store and retrieve data | www.microsoft.com /en-us/sql-server/default .aspx |
| PostgreSQL* | An open-source, object-relational database system | www.postgresql.org |

**Bibliographic tools**

| | | |
|---|---|---|
| EndNote | A commercial reference-management software package, used to manage bibliographies and references | https://endnote.com |
| Mendeley | A desktop and web program for managing and sharing research papers, discovering research data, and collaborating online | www.mendeley.com |
| Zotero* | An open-source reference-management system | www.zotero.org |

**Data storage/backup tools**

| | | |
|---|---|---|
| Amazon Cloud | A web storage application advertised as providing a large computing capacity; faster and cheaper than building a physical server farm | www.amazon.com /clouddrive |
| Dropbox* | A free service that lets you bring your photos, docs, and videos anywhere and share them easily | www.dropbox.com |

*(Continued)*

TABLE 10.1  *(Continued)*

| Tool by category | Description | URL |
| --- | --- | --- |
| Evernote | Evernote makes it easy to remember things big and small from your everyday life using your computer, phone, tablet, and the web | https://evernote.com |
| Google Drive* | A file storage and synchronization service that enables user cloud storage, file sharing, and collaborative editing | www.google.com/intl/en/drive |
| **Statistical analysis tools** | | |
| R* | A free, integrated suite of software facilities for data manipulation, calculation, and graphical display | www.r-project.org |
| S-Plus | A statistical and data-analysis package that allows nonprogramming users to access advanced visualization and modern analysis | www.solutionmetrics.com |
| SPSS | Statistical Package for the Social Sciences, widely used in the social sciences | www.ibm.com/software/analytics/spss |
| Systat | A statistics and statistical graphics software package | https://systatsoftware.com |
| **Data visualization tools** | | |
| Adobe Photoshop | A raster graphics editor software package | www.adobe.com/products/photoshop.html |
| ArcGIS | A platform for designing solutions through the application of geographic knowledge | www.esri.com/software/arcgis |
| ArcGIS Online | Everything you need to create interactive web maps and apps that you can share with anyone | www.esri.com/en-us/arcgis/products/arcgis-online/overview |
| ArcGIS StoryMaps | Interactive maps and multimedia that make it easy to harness the power of maps to tell your stories | https://storymaps.arcgis.com |
| MapServer* | An open-source platform for publishing spatial data and interactive mapping applications to the web, originally developed in the mid-1990s at the University of Minnesota | https://mapserver.org |

Note: This is by no means a comprehensive list. Please consult the DataONE software tools catalog (www.dataone.org/software_tools_catalog) for additional resources and search the Internet often, as new tools appear regularly.

2007, Newman et al. 2011, Sheppard et al. 2014). Your challenge is selecting the right tools for you and your project and making these decisions early on. New tools are developed daily, so search for and evaluate them often.

When collecting data, you make decisions about experimental design that influence the data you generate and ultimately manage. Proper planning for data collection can include choice of experimental design (e.g., random sampling, stratified random sampling, and systematic, to name a few), choice of data-collection tools (field data sheets with pencil and notebooks, mobile apps, wireless sensor networks, etc.), and data-collection protocols and methods. The key is to think through these choices carefully, take time to consult statistical advice when needed, and document your decisions in detail (a point we will return to later in this chapter).

### Assure

Providing assurances about the quality of your data requires good documentation, proper experimental designs suited to your research questions, and rigorous data-collection methods and techniques. Collecting your data following a well-written standard operating procedure (often just called an SOP) or protocol, using rigorous instrumentation, and documenting each of these aspects will provide assurances to those wishing to reuse the data for further analyses. Ensuring data quality requires meeting the criteria necessary for your research goals while adhering to relevant scientific standards. A good way to determine appropriate quality criteria is to work through your desired analyses and displays before any data are collected, particularly for citizen science projects that may have multiple contributors to a single database—a situation that makes later troubleshooting more complex and challenging. Using small test datasets created to include known potential problems (e.g., missing values, suspicious time/date values, and unlikely locations) is helpful for identifying ways that data-quality issues can be detected, corrected, and prevented (Wiggins et al. 2013). The approaches and mechanisms you use to ensure data quality apply before, during, and after data collection. Approaches applied before and during data collection are referred to as quality assurance (QA), whereas those applied after data collection are called quality control (QC).

*Quality assurance* refers to processes used to ensure that the best possible data are collected. In citizen science, QA is strongly linked to the design of participation tasks and supporting technologies. *Quality control* refers to a set of processes for evaluating the quality of data after they have been collected (e.g., data cleaning and decisions about how to handle missing data). QA/QC processes vary substantially in cost and effectiveness, but QC is generally considered more difficult and resource-intensive than QA (for a more detailed discussion of data-quality mechanisms commonly used in citizen science, see Wiggins et al. 2011). In projects dealing with visual processing tasks, such as transcription and image classification, having data entered twice by different volunteers is both feasible and a robust approach to ensuring quality and correctness. Many projects also collect paper data sheets, in addition to online data entry, for reference if and when questions arise about the accuracy of data entry. Similarly, voucher specimens can be collected and stored to serve as a reference point for taxonomic identification where such specimens are allowed and permissible.

### Describe: Metadata

Documenting your data requires recording the data *about* your data or metadata. Metadata include things that may affect the quality, accuracy, or precision of your data or the context or conditions under which they were collected. Examples of contexts and conditions include the weather during sampling, the instrument make and model used to record or collect the data, and the protocol followed, along with descriptive information about the type of data and their format and meaning. Metadata can also include descriptive information such as a short title and abstract communicating your study's intent (figure 10.3) or detailed information about a given column in a dataset generated by your citizen science project. Documenting as much metadata as possible is a good rule of thumb. Recording who collected each observation and its date, coordinates, units, and the model of the device used keeps metadata closely attached to each measurement, allowing the metadata to vary record-by-record. However, things like taxonomic specificity or weather conditions would be redundant if listed in each row, so including those metadata in an associated README file might be more appropriate. Some software automatically records useful metadata on your behalf. For example, the geospatial software ArcGIS records metadata related to transformations made, formats used, and spatial attributes associated with spatial data such as datum, projection, and coordinate system. Regardless of the storage approach for your metadata, the key is to make sure you record it all.

Given that many items could be considered metadata, it is valuable to decide ahead of time what will be recorded. When deciding, be sure to consider your project, protocol, data sheet, QA/QC procedures, and volunteer training. Document as much detail as possible about your research process, steps, methods, and data. This documentation is important for two key reasons. First, you want someone to be able to pick up with the research and continue in the future where you left off should you leave. Second, you want others to know exactly what you did so that your data can be integrated with other data or be used to repeat your research. When reporting metadata, it is critical that you write it all down. *How* you write it down is less critical—use a text file, a word processing file, a voice recorder, a PowerPoint diagram, a field notebook, or even a napkin in a pinch. If you want your metadata to be machine readable, use a standard metadata format (e.g., Ecological Metadata Language; table 10.1). But do take time to write it down. Keeping information in your head and assuming that you will remember it later (or that you will be around five years from now) will inevitably backfire. Many valuable data collected by individuals (professionals and citizen science practitioners alike) over the years have failed to record any metadata, which can ultimately lead to data loss (or death).

### Preserve: Data Storage and Backup

Once you have compiled your data into a data package—including good documentation and metadata—you need to decide where to store and preserve it. *Storage* refers to the mere saving of data on hardware resources. Storing and backing up citizen science data is a key aspect of data management. After all, you went to the trouble of collecting valuable, high-quality, and rigorous data, and you would hate to lose it. When making data storage and backup decisions, consider the following:

How much data do you anticipate collecting?

What hardware resources do you have?

How will you ensure that data are stored in separate locations (data redundancy)?

Have you written a README file to describe your files?

What file and folder naming conventions did you use?

Have you documented your conventions?

What type of versioning did you employ?

Data *preservation* usually involves bundling all related files associated with a research project into a dataset, a series of datasets, or a data package. A general practice we find useful is to include a README file for your data package that describes and labels each file and dataset such that a researcher independent of the project could read it and understand the title and contents of each file. Preservation takes into account the support services associated with storage, including file format migration (when applicable and needed) necessary for maintaining data over time. One way to accomplish data preservation is by using a data center or archiving service familiar with your area of research. These services can provide guidance on formal metadata preparation, data preservation, file formats, and how to provide additional services to future users of your data, including these important data preservation tips (Strasser et al. 2011):

Identify data with long-term value.

Store data with appropriate precision.

Use standard terminology.

Consider legal and other policies.

Document provenance and attribution details (e.g., person responsible, context of the data, revision history, links to source/raw data, project support/funding, proper citation, and intellectual property rights or licensing considerations).

Cloud resources such as GitHub, Amazon, Dropbox, Google Drive, and OneDrive (to name a few) offer good data storage opportunities but often lack preservation support. Other options for backing up data include external hard drives, portable flash drives, and local area networks that may be available, each with positives and negatives to consider (for additional storage and backup options, see table 10.1).

### Discover: Data Discovery, Accessibility, and Sharing

Once you have ensured that they are stored, backed up, well documented, and preserved, your data are poised for efficient discovery, accessibility, reuse, and sharing. Data discovery indicates how easy it is for others to find a dataset and reuse it. One step you can take to improve the discoverability and subsequent accessibility of your dataset is to store it, along with associated metadata, in a data repository or data center. Think of these as homes for data (table 10.1). Data repositories are often associated with a library, university, museum, scientific journal, or field station and are interested in (or often required to focus on) archiving and preserving information. These organizations have tools to package digital data and metadata in standard formats, migrate file formats over time, supply persistent identifiers for citations (i.e., digital object identifiers, or DOI), and keep sensitive information (e.g., endangered species locations) secure. For example, the Avian Knowledge Network (AKN) serves as a data repository for ornithological data, and, increasingly, journals like *PLoS ONE* and Data Papers in the journal *Ecology* require data deposition in their own repository as part of publication. Other repositories include the Knowledge Network for Biodiversity (KNB) and the Global Biodiversity Information Facility (GBIF). These data centers are skilled at data archiving and preservation. If you cannot find a repository, you might consider emailing your data files to a land manager as an alternative or check out GitHub as another approach. Putting your work out there in as many places as possible allows others to discover, access, and reuse it, especially when these places are machine discoverable through automated searches. Finally, sharing data satisfies funder expectations, increases the rate at which science can advance (Nielsen 2012), and saves resources. Citizen science itself is a form of data sharing, and the challenges of using citizen science data reflect classic data-sharing challenges (Hampton et al. 2013), such as how one can know whether to trust the data.

### Integrate: Data Standardization and Integration

Once your dataset is stored, how can you make it desirable for reuse? Standardizing data with other similar data enables syntheses that would not otherwise be possible. For example, one synthesis approach is meta-analysis, a method for combining and contrasting results from different studies to discern any consistent patterns, disagreement, and other relationships across a set of studies (Greenland and O'Rourke 2008). Similarly, systematic reviews and analyses are conducted across studies to evaluate patterns, relationships, and trends (Sutherland et al. 2004). As mentioned in chapter 9, an easy way to standardize data for integration is to use existing protocols. Furthermore, websites such as CitSci.org allow you to standardize data across projects easily (Newman et al. 2011). Specifically, in CitSci.org, project coordinators create their own measurements and then coordinators of other projects can select any of these created measurements for use in their own data sheets. Thus, projects can share measurements and standardize data such that they can report data with the same names and units to simplify data integration and synthesis. Using the same names or vocabularies for categorical data is another way to standardize. Similarly, standardizing definitions of terms is important to facilitate connecting databases. For example, the Nelson and Graves (2004) amphibian calling index classifies the degree of amphibian vocalization into several categories. Consistent use of these published categories by many projects is helpful, rather than having each project create its own categories.

Analyze

In citizen science, data are commonly transparent, open, and readily available. Volunteers can analyze data themselves just as project managers, natural resource managers, scientists, and policymakers can. Regardless of who is doing analysis, caution must be taken by all to ensure accurate, precise, and meaningful interpretations, results, and conclusions.

Data analysis is the process of manipulating data to yield useful information. To analyze something is to examine it methodically by separating it into parts and studying interrelationships and patterns. In particular, analysis—often statistical analysis—allows you to make inferences about possible answers to questions. For instance, "Is species richness different between forests, grasslands, and wetlands?" Your investigation hinges on your answer, which depends on your analysis. If your analysis indicates that diversity is indeed different between habitats, this result not only can inform basic science, but also may be used to aid in making science-based decisions. For example, if species richness were to differ between types of land cover, such a finding could provide information to a county planner on where (and where not) to locate roads. Data analysis is not always statistical, however, and there are a number of other approaches, such as qualitative analysis of interview data, analyzing pixels in images, assessing wavelengths in recordings, and analyzing interpretations made by volunteers classifying images. Some common analysis methods for categorical, time-series, and multimedia data typically collected in citizen science include statistical comparisons (categorical and continuous data), trend analyses (time series data), and transcription and qualitative coding (multimedia data).

Regardless of the analysis method, data must be properly prepared prior to carrying out any analysis. Data preparation is a way to begin the iterative process of data cleaning to ensure accurate and precise analyses. Preparing data involves checking for outliers, cleaning data (e.g., running QA/QC scripts), formatting data, and checking for and performing any iterative changes needed (e.g., transforming data). A more comprehensive discussion of preparatory methods can be found in Neil Salkind's (2011) entertaining book *Statistics for People Who (Think They) Hate Statistics,* but suffice it to say that a well-prepared spreadsheet will make subsequent analysis much easier and will generate more meaningful interpretations, results, and conclusions.

## DATA VISUALIZATION

Data visualization helps communicate discoveries. For example, as a practitioner with limited resources, you might create a visualization of the number of volunteer hours by year. Want to inform policy? Show your data with this goal in mind. Want to show results? Devise a chart to show how species richness differs between habitats. Now stakeholders can quickly see key points and apply results to decision making. Data visualizations can take many forms, from maps and charts to tables, animations, and infographics. Additionally, the data you choose to visualize can be raw, transformed, summary statistics, or some combination. You might choose to show monitoring locations on a map, pH measurements at station no. 1 over time, or the percent cover of noxious weeds at Fox Meadow Park. Data visualizations can be a snapshot in time or generated in real time, can be animated or still, and can take the form of pie charts, histograms, bar charts, line charts, scatter plots, box plots, or heat maps, to name a few options. In any case, data should be visualized to convey variability (e.g., standard error bars) if the data are statistical in nature, as well as axis labels and legends to ensure that your audience understands the information being presented. Ultimately, each visualization approach has pros and cons to navigate by considering your project objectives, stakeholders, and audience.

### Audience and Purpose

The first step to meaningful data visualizations is effective data management throughout the duration of your project. A second important step includes identifying your target audience and the purpose of your visualizations. You may find that you need to create several visualizations of the same data for different audiences. For example, you may need to create an engaging and professionally designed graphical figure for a newsletter that summarizes the annual progress of your citizen science project for your volunteer participants, but you may also need to generate a scientific table of results for a peer-reviewed journal article in collaboration with your partner scientists. The key is to identify the most appropriate visualization approach for your audience and purpose.

### Tables

Tables are a reliable and popular approach to visualization, allowing readers to compare and contrast data and make choices. Consider the information presented in figure 10.4 (table A) and note how the use of check marks signifies how a given citizen science approach empowers participation at various research steps. The use of icons and symbols makes it easy to compare approaches. Similarly, in figure 10.4 (table B) we can easily see that pH is increasing at plots 1–3 but decreasing at plots 4 and 5. Tables also organize and communicate large amounts of data in a small space while showing raw values. The disadvantages of tables include the difficulty for readers of identifying trends when numerical data are presented, the difficulty of printing large tables, and their lack of ability to present multimedia data.

### Maps

Always record locational data (latitude and longitude) while in the field. Armed with location data, you can make maps, which are great for showing spatial relationships and geographic differences or similarities. Benefits of maps include their ability to be meaningful to many audiences, show patterns, and provide context. A choropleth map (figure 10.5A) is a special type of thematic map in which areas are shaded in proportion to the measurement displayed. Choropleth maps provide an easy way to show how a measurement varies across a region. Likewise, animated maps can be useful for showing both spatial and temporal trends in data. Maps also easily show monitoring locations in relation to a region. When making any map, it is standard practice to include a title, scale bar, north arrow (or compass rose), projection/datum information, author, data sources, and the date the map was produced.

| Table A | Citizen Science Approach/Model | | |
| --- | --- | --- | --- |
| **Research Step** | Contributory | Collaborative | Cocreated |
| Define a question/issue | | | ✓ |
| Gather information | | | ✓ |
| Develop explanations | | ✓ | ✓ |
| Design data collection methods | | ✓ | ✓ |
| Collect samples | ✓ | ✓ | ✓ |
| Analyze samples | ✓ | ✓ | ✓ |
| Analyze data | | ✓ | ✓ |
| Interpret data/conclude | | | ✓ |
| Disseminate conclusions | | | ✓ |
| Discuss results/inquire further | | | ✓ |

| Table B | Plot # | 2005 | 2006 | 2007 | pH Trend |
| --- | --- | --- | --- | --- | --- |
| | 1 | 5.2 | 6.3 | 6.8 | ↑ |
| | 2 | 4.8 | 4.9 | 5.2 | ↑ |
| | 3 | 6.5 | 6.8 | 7.1 | ↑ |
| | 4 | 6.8 | 5.2 | 4.2 | ↓ |
| | 5 | 5.1 | 5.0 | 4.3 | ↓ |

FIGURE 10.4. Two examples of how tables can effectively visualize data. Table A: Volunteer participation in each of three approaches/models of citizen science projects (e.g., contributory, collaborative, and cocreated) within each of the 10 steps in a typical research process. Check marks signify participation in the specified step. Table B: Stream pH values reported in 5 plots over three years (2005, 2006 , and 2007) along with the associated trend in stream pH for each plot.

While maps are quite valuable, they can be difficult to create. Making a good map requires GIS or other cartographic software and advanced understanding of the software and associated spatial data. Sometimes good maps also require transfer of geospatial data from a GIS into a graphic design software package such as Photoshop. Finally, creating animated maps is difficult, presenting numbers spatially can be tricky, and displaying too many locations can make maps difficult to read.

### Graphs and Charts

The primary advantage of graphs and charts is that they help your audience visualize the point of the presentation. Graphs emphasize the main point. Specifically, graphs make data more convincing, provide a compact way of presenting information, and help audiences stay engaged. Design your graph with the end in mind. That is, what do you want the viewer to see and what is the purpose of showing your data with the graph or chart type you have

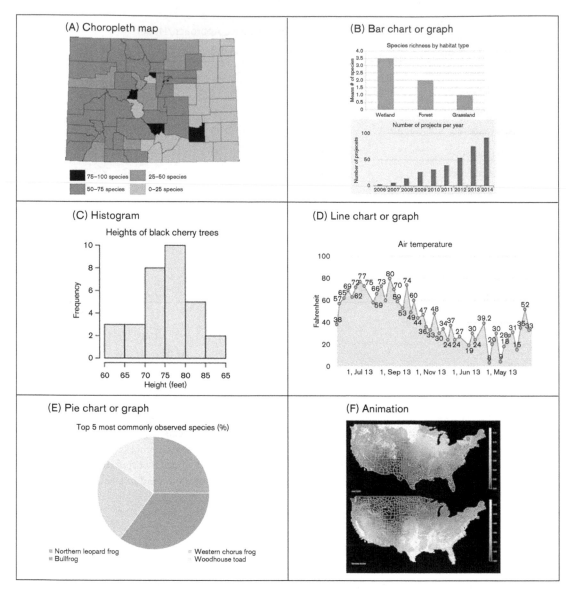

FIGURE 10.5. Examples of visualizations, including (A) a choropleth map showing species richness, (B) a bar chart or graph showing species richness by habitat type and a cumulative bar chart showing number of projects per year that used CitSci.org, (C) a histogram showing frequency of black cherry tree height, (D) a line chart or graph showing air temperature at a Trout Unlimited stream-monitoring location, (E) a pie chart or graph showing the top five most commonly reported amphibians in a CitSci.org project, and (F) two time-lapse frames from an animation showing the migration of the Savannah sparrow in summer (top) and winter (bottom). Your choice of chart, graph, or map will depend on the type and volume of data you have and the take-home message(s) you aim to convey. Credit: data are from CitSci.org and eBird (hypothetical data also included); images in panel F are from eBird (Kelling et al. 2009).

chosen? The answers to these questions often lie in the type of data you have, the volume of data, and the message(s) you wish to convey.

Citizen science data can be visualized using a variety of graphs and charts, each of which has its own advantages and disadvantages. A *bar chart or graph* uses horizontal or vertical bars to show comparisons among categories (figure 10.5B). One axis of the bar chart shows the categories being compared and the other represents a discrete value. Some bar graphs present bars in groups while others show bars divided into subparts for cumulative effect (stacked bar graphs). A *histogram* is a special bar chart presenting the distribution of data and represents an estimate of the probability distribution of a continuous variable rather than a discrete response (figure 10.5C). Histograms show the frequency of each response observed and are useful for inspecting normality. A *line chart or graph* displays information as a series of points connected by straight lines (figure 10.5D). Line charts are similar to *scatter plots* except that their points are ordered. Line charts show how particular data change at equal intervals of time (Salkind 2011). A *pie chart or graph* is a circular chart divided into sections (slices) to show numerical proportions (figure 10.5E). Variants include the exploded pie chart, the polar area diagram, the ring/sunburst/multilevel pie chart, and the doughnut chart. Problems arise in discerning various percentages of slices when there are too many slices or when the amount of a given slice is small in relation to the whole, but these charts excel when slices are large and few in number. Finally, *time-series graphs* display data at different points in time and are often quite useful in citizen science contexts.

Disadvantages of graphs include being time-consuming to construct and costly to produce. Often, graphs and charts require specialized technology that many citizen science practitioners do not have easy access to or cannot afford. Hence, it can be helpful to reach out to nearby universities or science centers (or even federal agencies) to see if these organizations have graphing software, technologies, or expertise to offer.

### Animations

Animations bring your topic to life. In particular, animations grab the attention of your audience and can be fun to watch. Technically speaking, animations create an illusion of continuous motion by rapidly displaying a sequence of subtly changing images. Among the best examples of citizen science animations are those developed by eBird (Kelling et al. 2009) to depict predicted bird migrations (e.g., of the Savannah sparrow; figure 10.5F).[2] You can create your own animations as well. Microsoft PowerPoint has easy-to-use tools that animate graphs or charts. Sophisticated software such as Photoshop, Fireworks, and ESRI ArcGIS help make animations that are more complex. For example, you can use the ArcGIS time slider tool to make animations of spatial data over time.[3]

2. For the complete animation, see https://ebird.org/content/ebird/wp-content/uploads /sites/55/SAVS_large.gif.

3. For a good tutorial, see www.youtube.com/watch?v = jkf86Ft1muA.

✓   Develop a data management plan.

   • Include sections for types of data, data formats, metadata, policies, protocols, storage, backup, and budget.

   • Identify a centralized data-management system for sharing and depositing data (i.e., individual workstation, cloud server, online system, or data repository).

   • Keep your plan simple.

   • Use free tools such as the DMPTool (https://dmptool.org), a widely accepted data-management tool in the ecological sciences.

   • Identify possible repositories that may be interested in preserving your data for the long term (see http://databib.org).

✓   Select appropriate software.

   • Use online websites if resources are limited (e.g., CitSci.org).

   • Use existing systems for preestablished projects (e.g., eBird or Project Budburst).

   • Use spreadsheets if you have a single, simple project of short duration.

   • Use a database if you manage more than one project over many years or if your data schema is more extensive.

✓   Employ consistency in managing files and data.

   • Add a date (YYYYMMDD) and timestamp (HMS) to everything (database tables, filenames, etc.).

   • Version all files and consider adding author initials to filenames (i.e., Version 1_NEK, Version 2_NEK, etc.).

   • Always include a unique ID for each row in all spreadsheets and database tables.

   • Identify meaningful naming conventions by using or creating controlled vocabularies when available (e.g., Upland Plot 1, Upland Plot 2).

   • At any level in the hierarchy of folders, subfolders, and files, keep contents consistent. We recommend not mixing subfolders and files; instead, have either all files or all subfolders.

   • In a relational database, use prefixes such as "TBL_" for tables, "REL_" for relationship tables, and "LKU_" for lookup tables to help organize similar tables by name.

   • Document your method of signifying missing data and how you distinguish between zeros and missing data.

✓   Create metadata.

   • Document minimum information, including a dataset title, creator, contact information, and keywords.

   • Write a descriptive abstract, which includes who, what, where, when, how, and why data were collected.

   • Describe or provide references for methods and/or experimental designs.

   • Create a list of attributes in the dataset, which includes column header name given to the measurement, description of the measurement, and unit of measurement.

   • Define codes when necessary (e.g., LABU = lark bunting, M/F = male/female).

   • Signify symbols used for missing data versus zero values.

Disadvantages of animations include the amount of effort and the skill in using animation software necessary to create them, as well as the need to access such software. Additionally, too many animations on the same web page can be distracting or even annoying to your audience, can consume a lot of bandwidth and slow down page load time, and often require special browser plug-ins to watch.

## GUIDELINES

Given our experiences in managing the complex data that arise from citizen science, we have compiled the following checklists (✓) along with associated guidelines (•) for data management (box 10.1), analysis (box 10.2), and visualization (box 10.3).

## A FEW NOTES ON NAMING CONVENTIONS

Naming conventions are consistent patterns used when naming plots, records, observations, database tables, or files. Defining naming conventions at the start of your project is extremely useful. For instance, you may want to consider how files will be ordered (if you want recent dates to show up first, consider using a convention like YYYY_MM_DD_Name.xls). Other things to consider include how you want to represent versions of files as changes are made. For example, you might use this convention: start with ReportName_v1.pdf, and for subsequent versions change v1 to v2 (etc.). Prefixes are also useful. Adding an underscore as a prefix on Windows systems helps bring important documents or files to the top of a folder when listing them alphabetically, regardless of the starting letter of the filename. Finally, for the database developer, using clear table names and prefixes—such as "TBL_" for tables, "REL_" for relationship tables, and "LKU_" for lookup tables—helps organize similar tables in a relational database and makes them easier to understand.

## SUMMARY

Data management and visualization are activities that allow you to share results emerging from the data collected by your citizen science project in meaningful ways. Through these activities, you can transform your work into tangible products and yield significant positive impacts and outcomes. Bringing your own creativity to each of these aspects of citizen science, especially during data visualization, is an opportunity to elevate the impact of your program.

Citizen science often involves the collection of large amounts of data and it can sometimes rely on tasks such as mobilizing previously unavailable data, scaling data products, and using custom analytical tools. Such tasks bring with them inherent challenges in data validation, documentation, visualization, and storage, much as macro-ecological and big data approaches do. Robust data management must be part of any citizen science research project, and the full engagement of specially trained personnel is indispensable for success

✓   Use data-collection protocols geared toward desired analyses.
   • Make sure your protocols and sample size allow for the analysis you want and give you the necessary statistical power (if analyses are statistical in nature).
   • Make sure you know ahead of time what analyses you are planning to do and what data are needed for them.
✓   Consult a statistician early on in your project for advice (if your project is planning on statistical analyses).
   • Check local universities and Cooperative Extension offices.
   • Check in with your statistician often.
✓   Prepare your data before analysis.
   • Clean your data to remove outliers (but always maintain a copy of the original unclean dataset just in case).
   • Make sure that all data are in the same units.
   • Check for duplicate records.
✓   Keep analyses simple and topically relevant.
   • Use analyses that are meaningful to your project.
✓   Be mindful of significant figures and rounding.
   • Use a precision needed for analysis and that matches what was collected.
   • Avoid showing unnecessary precision in reports (one or two decimals at most).

(Rüegg et al. 2014). While training practitioners in data management is necessary, the inherent complexity of diverse tools and products necessitates that skilled professionals be included on teams (Rüegg et al. 2014), yet this can be too expensive for many citizen science efforts. Publishing data and aptly rewarding team participants for such publication should be part of your definition of success and supported by funding agencies and institutions (Goring et al. 2014). Data documentation and preservation are critical stages of data management, because scientific data are irreplaceable and will likely become part of a future integrated project or provide the basis for new research and discoveries (Rüegg et al. 2014). Hence, citizen science practitioners must integrate data management into project design at inception and throughout implementation. Such efforts will lead to improved communication and sharing of knowledge among diverse project participants, better science outcomes, and more transparent and accessible science (Rüegg et al. 2014). Ultimately, science-based decision making may hinge on the techniques you have chosen and applied for data management, analysis, and visualization.

BOX 10.3    Data Visualization: Checklist and Guidelines

✓    Identify your primary take-home message.

　　• Write down your message before making your visual product.

✓    Keep your audience in mind.

　　• Avoid using colors that cannot be distinguished by color-blind viewers (e.g., red-green).

　　• Make sure visuals can be printed in black and white easily and clearly.

✓    Choose your visual appropriately.

　　• Choose maps for showing spatial patterns.

　　• Avoid pie charts that have numerous categories or slices with very small values.

　　• Use stacked bar charts to show cumulative effect.

　　• Use scatter plots if you have thousands of data points.

✓    Always show measures of variability.

　　• Show error bars, standard deviations, variance, and confidence intervals.

　　• Show variability for maps as well, using heat maps of error when possible.

　　• Keep line chart × axis values equidistant when possible.

✓    Keep visualizations simple and focused.

　　• Avoid showing too much information on any one visual.

　　• Follow the "one visual, one message" rule.

✓    Always include important metadata about your visualization.

　　• Include title, author, date, data sources, projection/datum info, scale bars, legend, and north arrows on all maps.

　　• Include title, author, date, data sources, $x$- and $y$-axis titles, and legend for other chart types.

# ELEVEN

# Reporting Citizen Science Findings

EVA J. LEWANDOWSKI and KAREN OBERHAUSER

Reporting the findings of your scientific research is essential. Only through sharing the results of your project with others can it have an impact on the world around you. As a citizen science practitioner, you may already have experience sharing the results of traditional science (i.e., that conducted by scientists associated with an academic or research institution) or you may be new to reporting scientific outcomes. In either case, the focus of this chapter is on best practices for reporting the results of your project to others. Some of these best practices will be similar to those used in reporting traditional scientific research, while others are based on more general communication strategies or are unique to citizen science. As you consider reporting the findings of your project, it is valuable to keep in mind the following questions:

What goals do you have for sharing your research?

What are the main messages that you want to share?

With what audiences are you most interested in sharing your research?

What reporting media or venues do you have at your disposal?

Are there technological, monetary, or people-power limitations that you need to consider when deciding how to share your research?

When establishing a plan for reporting your research, you first need to determine the audiences with whom you wish to share your results. Here, we focus our attention on the following audiences: professional scientists, citizen science volunteers (for information on what to call volunteers, see box 11.1), the general public, and policymakers. These four groups

What do you call the people who contribute to your citizen science project: volunteers, participants, citizen scientists, or something else completely? Each project uses its own terminology, and different people will prefer to be labeled differently. For instance, some people may find the term *participant* too passive, while others may see *citizen scientist* as troublesome to those for whom political citizenship is a barrier. To decide what term you should use for your project, ask your volunteers. If they prefer one term over another, then you should honor that. When describing your project to other people, use the term your volunteers prefer, but be sure to clarify what that term means. Not everyone will know what you mean by "participant" or "volunteer," let alone "citizen scientist."

are the most common audiences for citizen science findings, but some projects prioritize reporting to additional groups, such as funding sources or the owners of land used for monitoring. While we do not specifically address the news media as a target audience, engaging the media can help you reach all audiences, as our examples will illustrate. Nurturing strong and positive contacts with the media often benefits reporting efforts.

After determining the audiences to target during reporting, you should establish goals for each audience. These goals may be general, such as increasing retention among project volunteers, or they may be highly specific, such as educating 20 percent of local landowners about the effects of their land-use activities on breeding birds. Rather than identifying target audiences and then creating reporting goals for each of them, you may find it more useful to first establish outreach goals for the entire project and then determine what audiences need to be reached in order to meet those goals.

When audiences and goals have been created, you need to decide what specific content will be included in reporting materials and the venues that will be used to share those materials. In many cases, content and venues can be used to reach multiple audiences. For instance, a "Latest Findings" section on a project website would be appropriate for reporting to both volunteers and members of the general public. Peer-reviewed journal articles (a standard means of publishing scientific results, in which research is reviewed by other professional scientists in the same field to ensure high quality prior to publishing), on the other hand, can be shared with volunteers but will be most appealing and effective for professional scientists.

Reporting is often considered the last step in the scientific process, but in citizen science it is more common for reporting to occur throughout the course of the project. Many citizen science projects involve long-term monitoring, meaning there is not a clear end to the investigation. As a result, waiting for the end of your project can lead to never sharing any results. Some project managers may choose to have all reporting coincide with the publication of a peer-reviewed paper, but in many cases, reporting yearly, seasonally, or even weekly can be more effective for keeping your target audience up-to-date. Regular communication is especially important when you are sharing findings with project volunteers and members of the public, who are often interested in the most up-to-date results available. Notably, frequent in-depth reporting requires frequent data analysis, and the ability to produce regular summaries of project findings should be considered as data-management systems are set up. Preliminary reporting can also have a downside, as we will discuss later.

## REPORTING TO PROFESSIONAL SCIENTISTS

Professional scientists[1] generally present the findings of their research in peer-reviewed publications or in other traditional forms of reporting such as presentations at professional scientific meetings. As is the case with most peer-reviewed scientific publications, a key goal of communicating citizen science research to the scientific community is sharing new and important findings and contributing to broader knowledge about how the world works. Citizen science has the potential to make substantial contributions to our understanding of the world when the data are analyzed and published. But, for better or for worse, there may be lingering skepticism about the ability of volunteers to collect valid data (Riesch and Potter 2014), and peer review can serve a secondary goal of addressing this skepticism on the part of scientists and broader audiences.

Highlighting citizen science findings in peer-reviewed articles not only serves to communicate the findings to other scientists, but may make them aware of the power and value of citizen science. As a result, publishing may encourage broader use of data collected by volunteers, which is sometimes lacking in conservation. For example, many citizen science datasets pertaining to monarch butterflies are underutilized, but in cases where monarch projects have engaged professional scientists, they have successfully analyzed these datasets and published the results (e.g., Davis 2015, Ries and Oberhauser 2015). However, that is not to say that citizen scientists cannot publish their results without the partnership of a professional scientist. Gayle Steffy, a long-time citizen scientist, recently published her research on over twenty years of monarch experiments and observations in a peer-reviewed journal, a feat she accomplished without a professional coauthor (Steffy 2015). Regardless of whether your project partners with professional scientists when publishing, you should always acknowledge the contributions of citizen science volunteers in your work. This acknowledgment may sound obvious, but in fact, many peer-reviewed journal articles neither clearly recognize the volunteers who contributed to their work nor even share the fact that volunteers were involved in the project (Cooper et al. 2015).

---

1. Typically we consider "professional scientists" to be people who are working in a scientific field, who have an advanced degree, and who are associated with an academic or research institution (e.g., government agency or nonprofit organization). However, there are professional scientists who fall outside of this classification, such as independent scientists.

Like presentations or peer-reviewed publications on other topics, reporting to scientists on citizen science outcomes should include clear summaries of new methods and results that indicate the scientific significance of the work and that are placed in the context of relevant existing research. Because of the potential skepticism of other scientists and reviewers regarding the ability of nonscientists to collect accurate data, reports should include detailed descriptions of quality assurance and quality control procedures (see chapter 10). These procedures can justify positive assumptions about the precision and accuracy of data that may have been collected with little help from professional scientists. Finally, because many citizen science projects attract participants on the basis of their societal relevance (e.g., conservation or human health), reports should include summaries of the management, conservation, or other impacts of the research whenever possible.

When considering publishing in a peer-reviewed journal, it is important to keep in mind that typically such journals are not widely accessible to participants. We suggest that peer-reviewed articles that use your citizen science data, along with all other content based on those data, be shared as widely as possible with broader audiences. One way to share your research widely is to publish in open access journals (e.g., *PLoS ONE* or *Ecosphere*), which are free for anyone to read. However, an important caveat about open access journals is that publishing in them can be costly (around $1,200–2,000 per article). Thus, if you plan to submit findings to an open access journal, budget for this at the start of the project (this is a necessary consideration even when submitting to traditional peer-reviewed journals, which often charge by the page to publish an article). Whether open access or not, in many cases the copyright agreement you sign with a journal will allow you to share some form of your published article or its results, either on your website or via email. Make use of these options as much as you can.

Publishing citizen science findings in traditional scientific venues can provide a strong foundation for communications to other audiences and should thus be a key goal of most citizen science projects. Research findings that have been subjected to peer review and published in a scientific journal will attract a broad audience much more easily than those disseminated in another way. Moreover, publishing in a peer-reviewed journal lends scientific credibility to the research, can clarify and improve the message, and can attract media attention. The publication process can also encourage project managers to use the data to their full potential and justify the time investment of the volunteers. Finally, sharing the release of a newly published journal article or advertising an upcoming scientific conference talk on social media is a common and popular way of informing others that you have results to share.

## REPORTING TO VOLUNTEERS

Reporting results of the research to your project volunteers serves several purposes, including increasing participant retention (see chapter 7), fostering knowledge, and encouraging behavior change. Furthermore, sharing results with participants is important because they want to know that their work is useful and contributing to a larger process (Rotman et al. 2012). In other words, cherish your volunteers by regularly updating them on the project's progress and results and acknowledging the importance of their efforts.

Many people join citizen science projects because they have a desire to learn more about the natural world (Bell et al. 2008), and sharing your project results is a perfect way to help participants learn. Citizen science projects are often designed with the intention of contributing to the education of volunteers, and in some cases that goal extends to encouraging conservation behaviors in participants. For example, butterfly citizen science projects like the Monarch Larva Monitoring Project, Journey North, and the Los Angeles Butterfly Survey make materials available to their participants that have an explicit focus on land management strategies, encouraging participants to engage in such activities as planting host and nectar plants for butterflies. Our own work suggests that sharing project-specific conservation information with volunteers is correlated with an increase in their participation in conservation (Lewandowski and Oberhauser 2017).

When reporting findings to participants, it is important to tailor the content to meet their needs. For instance, many citizen science projects are geared specifically toward children (Roy et al. 2012), making it important that content be designed for and made available to age-appropriate levels. Likewise, some projects involve participants who are non-English speakers, requiring that content be made available in other languages as needed.

When possible, peer-reviewed research findings should be made available to volunteers who want them. However, condensed, less technical versions of the peer-reviewed results are likely to be more accessible to most volunteers. In fact, the communication style used in peer-reviewed publications is often inaccessible to volunteers and to other important audiences (Gewin 2014), such as people who can influence policy. As a result, materials designed to share finalized project results with participants should be as brief and easy to understand as possible. Such materials should include an introduction to the overarching research question and summaries of the methods, key findings, and, when relevant, the ecological or conservation implications of the research. Moreover, these materials should highlight the role of your volunteers. For instance, did the volunteers come up with the main research question? Did they collect some of the data, and if so, which parts? Did they assist with analyzing or interpreting data? Your volunteers will want to see how they fit into the project, as well as what roles other volunteers are playing. You can indicate their roles and importance by including stories about individual contributions to the project or by summarizing the numbers and types of volunteers and their efforts.

Besides presenting the final or completed project results to your volunteers, you may also want to give them updates and ongoing results, including anecdotal accounts and preliminary results. Although reporting preliminary or late-breaking results can have benefits, there are also drawbacks. One drawback that can arise is when citizen scientists want to access preliminary results, but doing so could lead to inaccurate or misleading information being produced. As a result, professional scientists involved in the project may prefer to hold off reporting the

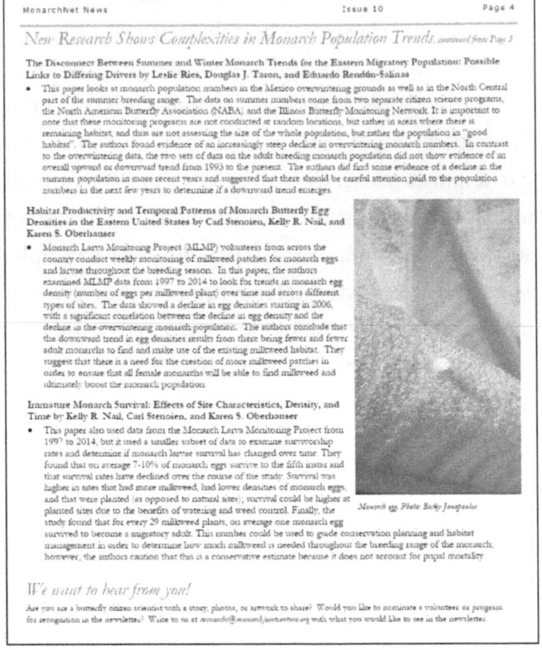
MonarchNet News Issue 10 Page 4

*New Research Shows Complexities in Monarch Population Trends, continued from Page 3*

**The Disconnect Between Summer and Winter Monarch Trends for the Eastern Migratory Population: Possible Links to Differing Drivers by Leslie Ries, Douglas J. Taron, and Eduardo Rendón-Salinas**

- This paper looks at monarch population numbers in the Mexico overwintering grounds as well as in the North Central part of the summer breeding range. The data on summer numbers come from two separate citizen science programs, the North American Butterfly Association (NABA) and the Illinois Butterfly Monitoring Network. It is important to note that these monitoring programs are not conducted at random locations, but rather in areas where there is remaining habitat, and thus are not assessing the size of the whole population, but rather the population in "good habitat". The authors found evidence of an increasingly steep decline in overwintering monarch numbers. In contrast to the overwintering data, the two sets of data on the adult breeding monarch population did not show evidence of an overall upward or downward trend from 1993 to the present. The authors did find some evidence of a decline in the summer population in more recent years and suggested that there should be careful attention paid to the population numbers in the next few years to determine if a downward trend emerges.

**Habitat Productivity and Temporal Patterns of Monarch Butterfly Egg Densities in the Eastern United States by Carl Stenoien, Kelly R. Nail, and Karen S. Oberhauser**

- Monarch Larva Monitoring Project (MLMP) volunteers from across the country conduct weekly monitoring of milkweed patches for monarch eggs and larvae throughout the breeding season. In this paper, the authors examined MLMP data from 1997 to 2014 to look for trends in monarch egg density (number of eggs per milkweed plant) over time and across different types of sites. The data showed a decline in egg densities starting in 2006, with a significant correlation between the decline in egg density and the decline in the overwintering monarch population. The authors conclude that the downward trend in egg densities results from there being fewer and fewer adult monarchs to find and make use of the existing milkweed habitat. They suggest that there is a need for the creation of more milkweed patches in order to ensure that all female monarchs will be able to find milkweed and ultimately boost the monarch population.

**Immature Monarch Survival: Effects of Site Characteristics, Density, and Time by Kelly R. Nail, Carl Stenoien, and Karen S. Oberhauser**

- This paper also used data from the Monarch Larva Monitoring Project from 1997 to 2014, but it used a smaller subset of data to examine survivorship rates and determine if monarch larvae survival has changed over time. They found that on average 7-10% of monarch eggs survive to the fifth instar and that survival rates have declined over the course of the study. Survival was higher in sites that had more milkweed, had lower densities of monarch eggs, and that were planted (as opposed to natural sites); survival could be higher at planted sites due to the benefits of watering and weed control. Finally, the study found that for every 29 milkweed plants, on average one monarch egg survived to become a migratory adult. This number could be used to guide conservation planning and habitat management in order to determine how much milkweed is needed throughout the breeding range of the monarch; however, the authors caution that this is a conservative estimate because it does not account for pupal mortality.

*Monarch egg. Photo: Becky Janopoulos*

*We want to hear from you!*

Are you are a butterfly citizen scientist with a story, photos, or artwork to share? Would you like to nominate a volunteer or program for recognition in the newsletter? Write to us at *monarchs@monarchjointventure.org* with what you would like to see in the newsletter.

FIGURE 11.1. *MonarchNet News,* a butterfly citizen science newsletter, regularly provides readers with brief summaries of recent peer-reviewed publications based on citizen science data.

research until it has gone through peer review. Such a delay can lead to tension between the professionals, who want to wait to share findings, and the citizen scientists, who desire more prompt reporting (Hoover 2016). Regardless of how often or at what stage you decide to share results, it is essential that you communicate your intentions before citizen science volunteers become actively engaged in the project. Being up-front with volunteers will allow them to make an informed decision to participate or not. If you choose to report preliminary findings, make it clear that these may not be representative of the final results.

A number of options are available for reporting your results to volunteers. Because most citizen science projects have an online presence, their websites are an ideal location for many types of reporting. Specifically, results can be placed as text on the homepage or other parts of the site (news and updates, blog sections, etc.) and/or can be made available as downloadable pdf files. Furthermore, you can send emails to project participants and make posts on social media to announce the addition of new results to the website. Social media posts are also an appropriate venue for sharing small but interesting updates, such as a record-setting observation, a species range shift, or the first sighting of the year of a species or phenomenon. Your social media (e.g., Twitter and Facebook) posts will garner more attention if they contain relevant photos, maps, or other graphics. In our work with butterfly citizen science volunteers, the majority of participants prefer to receive communications via email or the project website (Lewandowski and Oberhauser 2017). However, while many individuals have Internet access, not all have access or know how to use it. For instance, in the United States, 42 percent of people over the age of sixty-four do not use the Internet (Perrin and Duggan 2015). Thus, even if the majority of participants prefer online dissemination of material, it is important to have hard copies available.

Print or online newsletters are used by many organizations to share news and updates with their volunteers (Roy et al. 2012). Newsletters are an appropriate venue for content ranging from a short paragraph to about two pages. Short summaries of recent papers published by your project are perfect to share in newsletters (figure 11.1). If a project performs a good deal of in-person outreach with its volunteers, such as at training sessions (chapter 8) or educational workshops, single-sheet handouts can be an effective method of reporting results to volunteers. In-person events are also a prime location for delivering presentations that share your project results. These presentations should be accompanied by visual aids and can be supplemented with handouts.

## REPORTING TO THE PUBLIC

Whenever the public's attention is drawn to your citizen science project, there is potential to both share your findings and reach out to new volunteers and donors. Emphasizing the role of citizen scientists in your project will make it clear that volunteers are an essential part of the research, thus demonstrating that the project is a viable, worthwhile time commitment for its participants. Moreover, people who do not have the time or inclination to volunteer may feel more motivated to make monetary donations to a project that relies heavily on volunteers. Potential financial donors may also choose to give to projects that clearly describe the importance of their research, what results have already been achieved, and the conservation and ecological implications of those results. Similar to reporting project results to participants, another goal of reporting to the general public is often to create a heightened understanding of an ecological issue and to foster pro-conservation behaviors.

When reporting to the general public, your materials will usually require a bit more introduction and context than is needed in reports to your volunteers. In fact, you should assume that your audience will have little or no prior knowledge of the citizen science project or topic. In order to educate the public about ecology and conservation, the report should clearly describe what questions your project asks, what answers have been generated to date, and how those results factor into a larger ecological or conservation picture. Describing what actions are recommended for members of the public based on the new results is crucial for inducing behavior change among the audience. As with reporting to project volunteers, your reports to the general public can be done throughout the duration of the project and do not need to wait until final results have been obtained. However, you should take more caution when sharing preliminary results, as members of the public are less likely than your project volunteers to remain connected to your project and hear future, finalized results. As noted above, when presenting materials to the general public it is valuable to highlight the role of your volunteers, especially by including human interest stories, as this can increase participant recruitment and financial donations—and if the story is powerful enough, it can induce people to change their own behaviors. Notably, interviewing or describing the contributions of a local participant can also help make a local audience feel more connected to the story.

Television, radio, newspapers, and online news outlets all provide a critical and highly effective means of sharing your results with the public. In particular, these media outlets can provide a venue for scientists to make explicit connections between a current topic of interest and citizen science. For example, the Canadian Broadcasting Corporation interviewed one of us and three of our volunteers about the decline in monarch butterfly numbers, and the reporter emphasized the role that citizen scientists played in documenting and addressing this decline. News media can deliver results directly and direct their audiences to additional sources of information, such as a project website or upcoming event. In order to garner media attention, you can hire a publicist or create press releases and initiate media contact on your own. In either case, establishing a strong working relationship with members of the media, especially local media, is a good way to ensure continued coverage of your project.

Many of the same online venues used to recruit, train, and inform participants about a project (chapters 6–8) can also be used when reporting findings to the public. Project websites, blogs, and social media pages that contain your results should be accessible to members of the public and not require registration or login information prior to viewing. Sharing the links to these online pages through media, conservation organizations, or other citizen science projects will increase access and exposure to your project's results.

Finally, in-person reporting of project results at community events can be a highly effective way to reach the public, especially if your project is local. Presentations at libraries, town board meetings, fairs, and expos can be used to share project results. In some cases, you might host your own events and invite members of the public to learn about your findings. Public talks and workshops can include handouts that describe the project and its recent results.

## REPORTING TO POLICYMAKERS AND LAND MANAGERS

Many citizen science projects include conservation in their goals or mission statement, and many are focused on understanding impacts of climate change, detecting or managing invasive species, or understanding how habitat features affect species populations. If your project addresses a conservation-related topic, one of your goals will be to ensure that your results are communicated to people and institutions that can use the findings for conservation benefits, including maintenance or creation of science-based conservation policies and management strategies. In some cases, your volunteers may be working closely with the managers of the land they are monitoring or may themselves be managing that land, so many of the communication strategies outlined above for volunteers are relevant for this audience.

The feeling that "facts should speak for themselves" and that scientists should not advocate policy change still persists among some scientists (for a summary of this argument, see Meyer et al. 2010). However, a growing number of scientists, particularly those in conservation and other applied sciences, argue that scientists can and should be advocates (Meyer et al. 2010), and this argument translates readily into citizen science findings. In fact, the policy and conservation implications of citizen science findings should be clearly specified for policymakers (Gewin 2014). Land managers and policymakers are usually very busy, and they need clear translations of the specific recommendations that can be gleaned from citizen science data. As an example, the Monarch Joint Venture has used findings from several citizen science projects focused on monarchs to create a Monarch Breeding Habitat Assessment Tool (Caldwell and Oberhauser 2013) that includes suggestions on how habitat can be made more monarch friendly.

Management and policy recommendations can be communicated in many ways. As with other audiences, the status conferred by peer-reviewed publications, and any resulting publicity from those publications, can elicit interest on the part of land managers or policymakers. You can capitalize on this interest by providing land managers or policymakers with more detailed information, using summaries, concise bullet points of recommendations, or even presentations and one-on-one conversations. The key is to meet the goals of your own project while understanding and meeting the needs of other stakeholders as well.

## SUMMARY

Sharing your citizen science findings is an important step that should be built into your project's staffing, funding, data-management, and communication plans. Just as with data from any research, citizen science data that languish in drawers or on rarely visited websites are of little use (see chapter 10). The broad spatial and temporal distribution of many citizen science datasets make them especially important for conservation planning, and unless scientists, land managers, conservation professionals, and the public know the extent and value of the data, we run the risk of not taking full advantage of this conservation opportunity. A recent collaborative effort between the Nature Conservancy, rice farmers in California's Central Valley, Point Blue Conservation Science, and the Cornell Lab of Ornithology's citizen science project eBird illustrates the value of extensive citizen science data and knowledge of the availability of those data. The Nature Conservancy's scientists realized that providing habitat for migrating shorebirds could not be achieved by purchasing land. But data from Point Blue could predict the timing and location of water availability, eBird data could predict the timing and spatial distribution of twenty-six shorebirds that migrate through the Central Valley, and the Nature Conservancy could compensate farmers for keeping water in their rice fields to provide habitat for shorebirds during their critical migratory flight through the region (Jenkins 2014). None of this would have been possible had the eBird data not been well organized, available, and known to outside scientists and conservation specialists.

Because the target audiences for citizen science projects are very broad, you should consider a variety of dissemination methods and message content (table 11.1). Regardless of the venue or the target audience, it is crucial that you carefully ensure that your messages are clear and accurate. Such clarity and accuracy are especially important when reporting preliminary or in-progress results, which have the potential for confusion and misinterpreta-

TABLE 11.1 Tips for Using Different Media to Communicate with Your Audiences

| Media | Professional scientists | Citizen science volunteers | The public | Policymakers and land managers |
|---|---|---|---|---|
| Publications | Publish in peer-reviewed journals, especially open access journals when feasible. | Create newsletters or reports specifically for your volunteers. | Create summaries of findings and project methods to hand out at public events, if relevant. | Reference peer-reviewed publications, or create targeted white papers. |
| Talks | Present your research at professional conferences. | Host annual or seasonal meetings to share results. | Give talks to local groups, such as those focused on birding, gardening, or volunteering. | Share local results at town meetings; focus on policy implications. |
| Website | Provide links to peer-reviewed articles. | Post detailed summaries of recent findings. | Ensure public access; do not require logins or registration to view results. | Keep website professional and up-to-date. |
| Social media | Share new publications or announce upcoming conference talks. | Share recent, short, and exciting observations and findings. | Keep accounts open; do not require group membership to view posts about findings. | Include findings that are relevant to policy in social media postings. |

tion. Anecdotal observations and preliminary reports should be clearly labeled as such, and it should be explained that they may not represent the entire project.

While much of the work of reporting results falls to the staff and leadership of a citizen science project, your volunteers can also serve as disseminators. Many citizen science volunteers are willing and able to assist with more aspects of the project than just data collection, and relying on these volunteers to perform education and outreach about project results can greatly increase reporting capacity. Projects that plan to utilize volunteer disseminators should include training or materials to support their reporting efforts. For example, the Monarch Larva Monitoring Project offers a handout advising volunteers on best practices for contacting the media (Monarch Larva Monitoring Project 2014).

Finally, it is essential that all reports of your results acknowledge the contributions of volunteers (see box 11.1). Citizen science is wholly dependent on the efforts of its volunteer participants, and it is only fair that those efforts be referenced when explaining the results of a project. Furthermore, describing the role of volunteers can aid in recruitment and retention and can provide scientists with a better understanding of the potential power of citizen science participants.

# TWELVE

# Program Evaluation

REBECCA JORDAN, AMANDA SORENSEN, and STEVEN GRAY

Citizen science programs result in numerous benefits (Bonney et al. 2009), including advancing conservation goals (e.g., Danielson et al. 2005, Couvet et al. 2008, Devictor 2010), encouraging environmental stewardship (e.g., Cooper et al. 2007), and promoting engagement of the public in a more open form of science (e.g., Wilderman et al. 2004). Given these positive results, one might ask how project managers and those who research citizen science support their claims. How do we know the impact of our programs and the extent to which we have met the goals we set during project development? Addressing these and other questions is the focus of program evaluation.

Evaluation is the systematic process of improving a program by assessing the strengths and weaknesses of its organization, personnel, policies, and products. The American Evaluation Association (www.eval.org) provides several useful resources. We suggest that the most critical elements of citizen science program design are defining goals and objectives, creating indicators of success, and determining effects (table 12.1). Evaluation of these elements will allow the designer to develop and improve the program's activities, target outcomes, and necessary assessments.

Developing programmatic goals and objectives (hereafter "goals") is an important first step when creating an evaluation protocol. Making goals explicit can help focus program resources on the necessary elements for program success. Goals can vary widely but tend to focus on efficiency in program processes (i.e., how the program is undertaken), outputs (products, activities, and deliverables), and outcomes (impacts on the people and the environment involved with the program). Whenever possible, these goals should be established at the outset of designing a project.

TABLE 12.1   Guide to Conducting an Evaluation

To be photocopied and used as a broad outline. We recommend using this with the online resources provided in this chapter.

| Considerations | Notes | Checklist |
|---|---|---|
| **Goals** | | |
| Describe your overall purpose for your project. | | ☐ Yes<br>☐ Getting There<br>☐ Not Yet |
| Break down your purpose into separate and specific goals.<br>*Ask yourself: Can these be broken down further?* | | ☐ Yes<br>☐ Getting There<br>☐ Not Yet |
| Determine if each of your goals is measurable.<br>*Ask yourself: Can you think of ways to measure whether your goal has been met?* | | ☐ Yes<br>☐ Getting There<br>☐ Not Yet |
| **Indicators** | | |
| Create indicators (i.e., ways you know that your goal has been met) | | ☐ Yes<br>☐ Getting There<br>☐ Not Yet |
| Check to ensure that each indicator adequately addresses your goal.<br>Ask yourself: Do your indicators cover all your goals? | | ☐ Yes<br>☐ Getting There<br>☐ Not Yet |
| Determine how you will gather information for your indicator.<br>*Ask yourself if your approach is practical and whether the information gathered will be reliable.* | | ☐ Yes<br>☐ Getting There<br>☐ Not Yet |
| **Determining effects** | | |
| Create a way to determine if the information you gather reflects a meaningful measure of your project goals. | | ☐ Yes<br>☐ Getting There<br>☐ Not Yet |
| Develop an approach to determine if the differences that you are measuring are meaningful or an artifact of natural variation. | | ☐ Yes<br>☐ Getting There<br>☐ Not Yet |
| Determine if your information suggests a change in the inputs, activities, or outputs of your project. | | ☐ Yes<br>☐ Getting There<br>☐ Not Yet |

When determining the goals of your program, it is important that you do not limit yourself. You can write as many goals during this stage of the evaluation as are needed. In making goal statements, try to be as precise as possible. You can always remove and revise later in the program. The acronym SMART (Specific, Measurable, Attainable, Relevant, and Time-Relevant; Doran 1981) can be quite helpful in developing goals that can be managed and gauged.

Once goals are established, it is critical to develop the measures by which they can be evaluated. These measures, often termed *indicators* for outcomes and *targets* for outputs, allow you to determine the extent to which you have developed an efficient process (e.g., cost-effective training, time efficiency), appropriately established target criteria (e.g., number of participants recruited, number of trainings completed, number of observations obtained, number of publications, management plan developed), and created reliable indicators to evaluate your long-term impacts (e.g., participant attitudes and perceptions, learning about scientific concepts and processes, development of skills, development of projects and products). The most difficult part of this process is to create metrics that are reasonable and reliable.

Using reliable metrics and appropriate methods to determine to what extent your program has met goals, developed outputs, and addressed impacts will help you plan the next steps, report to funders and other stakeholders, and share results. Note that we deliberately used the word *extent,* because it is best to assume that your program will have an effect. Therefore, it is not necessary to develop goals that are stated in an absolute, binary manner (i.e., "does or does not meet a goal"); rather, you can measure on a spectrum or scale the influence your program has had in key areas. We view program evaluation as a means to measure the distance between program end goals and what is currently happening in the program.

Consideration of goals, measures, and effects works best when they are *intentionally designed alongside the citizen science program.* Below, we explore the issues and opportunities associated with goals, indicators, and impacts, using a common framework for each that defines (1) considerations for data collection and analysis at each step, (2) the types of learning that might occur at each step, and (3) considerations for understanding the perceptions and actions of volunteers at each step.

## GOALS

Goals can be broadly classified into data, learning, and perceptions and actions. While it could be said that generating data is a common goal of all citizen science projects, goals regarding the uses, outcomes, and ownership of the data—that is, how these data can be used by individuals and institutions—can vary. It is important to note that not all projects need to attain all types of goals.

### Data

Citizen science has been responsible for the collection of a great deal of reliable data. For example, projects with long-term monitoring goals across large expanses (e.g., eBird, USA National Phenology Network, and Community Collaborative Rain, Hail, and Snow Network) have enabled detection of changes across large spatial and temporal scales (see, e.g., Lepczyk 2005). In addition, citizen science data have been generated to accurately address goals focused on measuring biodiversity change. For example, projects have found change in biodiversity because of climate change (e.g., Root et al. 2003), patterns of plant species invasions (e.g., Crall et al. 2011, Jordan et al. 2014), abundance in animal species invasions (Delaney et al. 2008), and wildlife disease spread (Dhondt et al. 1998) using citizen science data. In other cases, however, the data were unreliable, suggesting that goals were not met. In one recent study, volunteers had difficulty identifying the woolly adelgid (Fitzpatrick et al. 2009), which led to a systematic bias in the data generated by the project. Other studies have indicated that task complexity, age, and experience can affect a project's volunteer quality (e.g., Delaney et al. 2008, Schmeller et al. 2009, Farmer et al. 2014) and data quality (Pierce and Gutzwiller 2007). In some cases, these issues in attaining data goals may be attributed to simple misalignment between protocols of professionals and volunteers (e.g., Ottinger 2009). Researchers have begun to develop measures to deal with recurring data issues (e.g., Thornton and Leahy 2012).

### Learning

Citizen science projects pose ideal opportunities for volunteers to learn about the social and scientific context and practices of the project alongside professional scientists and managers (Trumbull et al. 2000, Brossard et al. 2005, Evans et al. 2005, Crall et al. 2013). When making learning a goal, it is important to consider the type of learning and its practicality. When designing the learning environment, you should consider the project's context. Projects range in the number of individuals involved, data to be collected, personnel available for training, and resources in terms of time and money. The design should accommodate these limitations and also be flexible to allow changes in response to what is learned through multiple iterations of the program.

Decisions on how to make changes can be informed by the learning goals, and vice versa. For example, if you are looking to collect large amounts of data over a short period, a large pool of participants may be necessary. Given the time available for each participant, the learning goals should be limited to what can realistically be covered and perhaps should be confined to the data-collection protocol. Other information may use valuable time and only be a distraction. If, however, repeated data collection is necessary over time, extending time spent in the learning experience may be beneficial.

Because the data-collection environment may vary, it may also be wise to promote in the learners an ability to make decisions beyond the protocol alone. This may mean that if volunteers are likely to encounter variable types of data-collection environments (e.g., different growth habits of plants, spotting birds in different environments, samples taken from different locations based on environmental conditions of the day), they will need training that allows them to recognize plants or birds in contexts they have never encountered before or to make decisions about when and how to sample without the guidance of an instructor. In other words, participants may need to think independently and critically. Therefore, from the outset, it is important to consider these likely scenarios and train volunteers in making decisions about experimental design, interpretation, or data communication. Training for these skills likely requires time for the volunteers to practice and make mistakes. The latter can be particularly detrimental to the quality of datasets, requiring careful planning of the learning experience.

Project designers may also want learners to become aware of issues, retain concepts that comprise the environmental, physical, and social issues, retain concepts related to those issues, broaden their scientific identity, conduct aspects of scientific investigations, increase motivation for action, change personal behavior, or change policy or management protocols. Once the broader project goals are established, it may be wise to work backward to isolate the essential elements of what is to be learned. It is also wise to consider learning from the perspective of the program leaders, who are also learners, and not solely from that of the participants.

Recall that all goals should be specific, clear, and *measurable*. It can be helpful in articulating learning goals to integrate the task with the concept that is to be learned. For example, individuals may be expected to demonstrate their understanding in writing or in multiple-choice explanations and descriptions of particular social, scientific, or socioscientific concepts (e.g., Brossard et al. 2005, Evans et al. 2005, Jordan et al. 2012, Crall et al. 2013). In other cases, scientific-process skill goals may involve demonstration through a project, written scenarios, or a presentation (e.g., scenarios; Jordan et al. 2011, Crall et al. 2013) or via an embedded performance-based evaluation scheme (Ballard 2008). Such goals should be limited to a reasonable scope given project resources.

### Perceptions and Actions

An often desired goal of many citizen science programs is to foster perceptual or behavioral changes in volunteers. Isolating whether a specific program affects perceptions and behaviors may be confounded by many individual and social factors (see, e.g., Stern 2000, Biel 2003, Meinhold and Malkus 2005), such as self-efficacy, personal values, and habits. The idea that, when individuals are given a chance to confront their ideas and actions, they will make changes after a citizen science project is perhaps naive. When such goals are desired it is important to put ideas in actionable and assessable terms. For example, the desire for individuals to recognize the role for personal action related to a particular environmental problem is not entirely assessable. In particular, recognition is somewhat hard to measure. Similar to designing the learning goals, consider measurable actions such as "describe in writing or presentation" or "evaluate through the creation of individual management plans." The latter are much more tenable (table 12.2). Also, keep in mind that behavior change does not tend to occur during short-term participation.

### INDICATORS

Once goals are identified, it is imperative to create reliable metrics that will provide evidence of the extent to which goals are being met. Because there is no perfect evaluation, program

TABLE 12.2  Sample Evaluation Scenario

A group of interested individuals is engaged in a citizen science monitoring program collecting data about invasive plants on local trails.

| Learning Goal | Measures | Category | Rubric | | |
|---|---|---|---|---|---|
| | | | Successful | Developing | Undeveloped |
| *Conceptual knowledge:* Describe how we define invasive species as separate from exotic species | Pre/post surveys on knowledge definition | Definition | Includes the recognition that invasive species cause undesired impacts and can be exotic or not | Includes the recognition that invasive species cause undesired impacts only | Unable to provide a complete definition |
| *Values and attitudes:* Identify personal values and how individual behavior influences the outcome | Modeling of behavioral influences in the system, individual reflection pieces of values/decision making, scenario discussions, etc. | Personal values | Individual is able to identify personal values and how these values might influence outcomes | Individual is able to identify personal values yet is unable to articulate how these influence outcomes | Individual is unable to identify personal values or articulate how outcomes are influenced |
| | | Personal decision making | Individual is able to clearly articulate the alternatives and the costs/benefits of each alternative; additionally, individual is able to discuss the costs of making an error | Individual is able to clearly articulate the alternatives and their costs/benefits but does not discuss the consequences of making an error | Individual does not articulate alternatives |
| *Scientific practices:* Generate a visual representation of data (e.g., a plot on graph paper) | Individual can generate representations to account for data/patterns | Skills | Individual decided to represent information in an appropriate manner | Some data are represented in a mostly appropriate manner, but data representation is inconsistent | Individual represented information in a manner that is unclear or inappropriate |

designers can feel qualified to take this step. Certainly, experts can be found at colleges and universities, and particularly (in the United States only) at land-grant institutions. Similar to goals, we can consider measures for data, learning, and perceptions and actions.

### Data

Because the generation of reliable data is a cornerstone of citizen science, efforts to ensure and measure data quality are essential. Consider the potential sources of error in the desired dataset and associated protocol before designing the protocol (Jordan et al. 2012). In any particular dataset, a certain amount of error can be tolerated depending on the analysis (i.e., the type of statistical test to be used). This error needs to be considered when thinking about the participant protocol. For example, in one of our citizen science projects, we sought to engage participants in spotting target invasive plant species while hiking. In this manner, participants would be successful if they identified a true negative (i.e., no invasive plants identified when invasive plants were not present) or a true positive (identifying invasive plants when they are present). Error in the dataset would come from false negatives (missing an invasive plant) or false positives (identifying the wrong plant). Therefore, participants needed to be trained to identify the targeted species and then find that species against a backdrop of other species. Given this situation, we designed an evaluation protocol for reliable data by first piloting the protocol to determine with what species and in what locations errors were most likely to occur. This piloting enabled the establishment of validation measures for a subset of the hiking trails (i.e., the likelihood of missing a plant). To further determine likely error rates, volunteers were asked to collect data from a single validation trail (in order to obtain the volunteer accuracy range) and to collect the first plant identified (to detect plant identification error). For example, volunteers were asked to hike the common trail first and were given the expert report after their initial hike to allow them to return to the site to improve their identification skills. Additionally, experts could be asked to conduct similar measures to determine the error rate of typical expert data (e.g., Crall et al. 2011). In another recent study, when data on banded snails were compared with expert data, citizens were shown to produce a reliable dataset (Silvertown et al. 2011, Worthington et al. 2011).

In certain cases, datasets are built in part by experts and in part by volunteers, and the volunteers' work may result in sampling biases. For example, volunteers may report data from the most convenient locations or times (e.g., Dickinson et al. 2010; Jordan et al. 2014, in which volunteers were restricted to on-trail observations only). Post hoc data treatment has been used to deal with anomalous reports or unusual variance (Kery et al. 2010, Snall et al. 2011).

### Learning

Evaluations need to be practicable and realistic. Being able to define a concept does not mean that an individual can evaluate how that concept might change over time, nor does providing a definition of the concept imply that a person can use the concept in action. By having the learning metrics (or assessment metrics) reflect the learning goals, the chances of gathering successful data are increased. Citizen science program designers are encouraged to be creative with assessment metrics and the use of tools such as rubrics in pursuing performance objectives (i.e., completing the actual task). Such objectives can better inform progress and final project reports.

If some or all of the learning goals reflect some element of retaining or elaborating specific concepts, an evaluation can contain items that are akin to test items that individuals may have experienced in their formal educational experiences. Responses can be compared with that of a typical expert. Other indicators, such as having volunteers rate their confidence in their response or provide evidence for it, can also be used.

If other learning goals focus on engagement with the actual practices of science, then it stands to reason that practice-oriented assessments must be developed. Some of these may be focused on data collecting, analysis, or reporting skills. These are likely best evaluated through written essays or, perhaps better, through the evaluation of volunteer performance of the skill, as alluded to above. Based on the latter, one may be tempted to measure mastery alone (i.e., they did or did not meet the goal). Mastery might do well in certain cases, but if the skill is a major feature of the program goals, then taking the extent of the mastery may be desirable. Determining the extent of mastery would entail creating a scale (e.g., nonexistent, developing, adequate, and well-mastered) and then describing the features of the task along with how much of the skill must be attained to meet each category of the scale. This evaluation instrument is often called a rubric (table 12.2). Quantitative or qualitative values can be attached to a rubric to enable different analyses.

### Perceptions and Actions

Measuring perceptions and actions can be quite a challenge. Regarding perception, there are many examples of validated instruments that project designers may want to consider for reporting purposes. For example, specific scales focused on science and environmental literacy can be found online at the PEAR Institute's Assessment Tools in Informal Science (ATIS) (see www.pearweb.org/atis; e.g., the Views About Science Survey, or VASS). Furthermore, should the program designer choose, the development of attitude-evaluation metrics can be as straightforward as Likert-scale items (rated on a symmetric scale; e.g., strongly agree to strongly disagree), which have been shown to be quite robust. Such Likert-type responses can then be used to measure reliability of responses using statistical software that is often freely available online.

Measuring actions such as engaging in an environmental cleanup or changing buying choices is difficult because it is impractical to follow individuals through their lives (although having regular prompts for them to fill out journal entries or web logs along the way may provide a nice approximation). In these cases, you may need to rely on self-report metrics. Self-report metrics can be quite helpful but can also be problematic because, for example, individuals may not be entirely aware of their actions and may not accurately recall past events (e.g., Baranowski 1985).

## IMPACTS

Beyond just reporting changes, citizen science leaders often need to discuss changes in terms of meaningful increases and decreases. This involves a program evaluation report that can provide insight into the ways that the program is or is not meeting its goals. Such insight requires careful inspection into each of the goals and subsequent metrics.

### Data

Fortunately for many, using metrics to determine the extent to which data are reliable may not be difficult. Reporting about such effects can be straightforward for practitioners accustomed to working with scientific data. In many cases, stakeholders (e.g., funding agencies and donors) want a sense of whether the data can be used to address the scientific, policy, or management question of interest. Providing metrics of volunteer accuracy, reliability, and precision can address the previous question, along with providing a sense of whether there were enough data to answer the research question. These metrics can be based on qualitative inspection, statistical analyses, or indices for which interpretation information is provided.

### Learning

Interpreting learning data can be challenging for practitioners trained primarily in the sciences. An important point to remember at this stage is that there needs to be alignment between the learning goals and the associated assessment metric. This alignment will enable the learner to attend to the appropriate material and will ensure that the educator is assessing what was actually taught versus what may have been implied but not directly instructed. Remember that multiple measures can be used and reported to paint an informative picture of the type of change in learning seen in the project. Furthermore, it is critical to realize that change is not necessarily meaningful but depends on what is learned, your audience, and the variance in your dataset. For example, scientists often distinguish what changes might be statistically significant as compared to changes that may or may not be meaningful or make sense. Doing so will help ensure enough power for detecting an effect and, as mentioned previously, a brief consult with an expert or the literature can be highly useful. In many cases gains in content knowledge are reported as an outcome of a citizen science project, but longer-term gains in thinking are much harder to measure. Therefore, the latter are rarely reported.

### Perceptions and Actions

Determining meaningful change in individuals or communities as a result of the project may be a difficult undertaking. Again, aspects of human belief and behavior are quite complicated. Thus, it is quite difficult to determine when a behavior has changed, especially if you are using measures that are correlated with a behavior instead of measuring that behavior. Language for explaining these effects in final reports might include qualifying words such as *likely, associated,* or *is expected,* rather than declaring that behavior change is happening. Projects can also measure change over time, which may allow for detection of actual behavior change.


## TAKEAWAYS

Much too often, project directors are not given the opportunity to learn about successes and failures in citizen science program evaluation. Because of a generally positive attitude about engaging the public in scientific research and a strong push to publish positive results, insights about programmatic shortcomings are often not provided in the peer-review or extension literature. Fortunately, resources such as CitSci.org and the Citizen Science Association (www.citizenscience.org) enable practitioners to learn from each other. In addition, tools such as DEVISE (Developing, Validating, and Implementing Situated Evaluation Instruments), which was created specifically for citizen science, are available. In the coming years, new tools for evaluation will be made available and, much like citizen science itself, will enable yet another layer of questions to be addressed by experts and nonexperts.

What may be most obvious is that the goals, measures, and reported effects examples featured in this chapter are constrained to the data and individual-based outcomes. There is, however, a move to measure broader, community-level impacts. Shirk et al. (2012) present ideas to help guide programs that are consistent with stakeholder needs.

Another perhaps obvious feature of this chapter is that we do not give adequate attention to programmatic goals such as understanding strengths and weaknesses of the programmatic process. If programs seek to iteratively meet goals, an exploration of the process is warranted. It is worth considering questions such as the following:

- To what extent are resources being used? These resources can include money (e.g., budget surpluses or shortfalls), personnel (e.g., Were individuals efficiently engaged? Were some individuals overworked or underworked at any time?), and time (e.g., Were activities finished on time? Was there a rush?). Think also about the space occupied and the materials used and ask yourself if these were resourceful or cost effective. Measures might include an analysis of hours and dollars spent on particular tasks that can then be aligned with participant enrollment. In most cases, these outcomes are likely to be used solely for programmatic improvement.

- To what extent are personnel contributing to specific goals? Are all of your goals being managed, and by whom? Are there backups in place for the roles that personnel are taking?

- Are there elements of the program that are underperforming, performing at level, or perhaps overperforming? In this case, it is important to consider why something is not working. Often there is a tendency to discontinue something that is not working without fully knowing why. Doing so may lead to further problems in the project. Sometimes an outside opinion can be helpful.

- Are participants satisfied with the experience and expected to stay engaged for longer-term outcomes? This is critical for any citizen science program: understand what motivates your participants, and design programs that align your programmatic needs with participant needs. This is covered more fully in chapter 13.

## CONCLUDING REMARKS

We posed an important question at the beginning of this chapter: How do we know the impact of our programs and the extent to which we have met the goals we set during project development? We hope that you conclude, as we do, that careful designation of goals, measures, and effects—not only of the scientific data but of the other program outcomes that focus on individuals, communities, and systems—will help provide the necessary evidence to address these questions. This evidence can serve to justify the need for the project and communicate its successes to potential participants and funders, among others. Often, evaluations can also show program directors the range of accomplishments that can occur within a project period. Therefore, while the process may seem daunting, many find their program evaluation to be as interesting, enlightening, and essential as the broader scientific question being addressed by their project.

# THIRTEEN

# How Participation in Citizen Science Projects Impacts Individuals

REBECCA CHRISTOFFEL

Citizen science is an interdisciplinary venture, and projects have educational and/or engagement goals as well as scientific goals. This chapter elucidates the ways in which individuals and communities can be changed via their participation in citizen science, including changes in awareness, content knowledge, knowledge of the scientific process, scientific skills (e.g., identification and use of specialized instruments), attitudes toward science and the scientific process, and behaviors (e.g., involvement in policymaking). As discussed in chapter 12, there are ways to measure changes in participants resulting from their involvement in your project.

## WHO PARTICIPATES?

First, let's review the information we have about who participates in citizen science and what groups are underrepresented. Such information may be helpful as you develop your own project, particularly if you have a goal of increasing the diversity of participants or are specifically working with an underrepresented community that has approached you about co-creating a citizen science program. As noted in chapters 6 and 7, there are many specific motivations for individuals' decisions to participate in citizen science projects.

Adults who participate in citizen science are often already interested in science (having been self-selected to participate) and in the study organism(s) or process (Smith et al. 2015), and they often already possess specific skills (Crabbe 2012) or knowledge (Overdest et al. 2004, Homayoun and Blair 2008). Affluent individuals are overrepresented in citizen science (Trumbull et al. 2000, Rohs et al. 2002), as are highly educated individuals (Overdest et al. 2004, Evans et al. 2005). In some instances, a majority of participants have previously been involved or are currently involved in other citizen science groups.

People less likely to get involved in citizen science projects include minorities (Trumbull et al. 2000, Brossard et al. 2005, Evans et al. 2005). Underrepresentation of urban participants and overrepresentation of participants from the suburbs have been documented in some projects (e.g., Rohs et al. 2002, Evans et al. 2005). While we have already discussed a number of barriers to participating in citizen science projects (see chapters 6 and 7), others include family resources and engagement (Evans et al. 2005), balancing work responsibilities with other activities (Jolly 2002), and the perception that the science does not address a community's most urgent issues or mesh with its values (Pandya 2012).

## HOW ARE PARTICIPANTS CHANGED THROUGH THEIR INVOLVEMENT?

Apparent changes in individuals and communities that result from involvement in citizen science projects depend, in some instances, on the type of project being undertaken and the roles and responsibilities it gives to its volunteer participants (see chapters 1 and 3).

### Awareness

Perhaps the most readily apparent change in people who participate in citizen science projects is in their awareness of the issues they are investigating (e.g., water quality; Nerbonne and Nelson 2004). Awareness can lead to interest in and actions taken to increase knowledge about a particular issue or process (e.g., awareness of backyard as habitat; Evans et al. 2005).

### Knowledge

The scientific literature provides ready examples of changes in knowledge associated with scientific facts (Brossard et al. 2005, Trumbull et al. 2005), a scientific discipline (Jolly 2002, Brossard et al. 2005, Fernandez-Gimenez et al. 2008), ecological functioning (Nerbonne and Nelson 2004, Jordan et al. 2011), and the scientific process and its methods (Kountoupes and Oberhauser 2012, Bonney et al. 2016). However, gaps remain. For example, participants may not gain an understanding of objectivity in the scientific process or learn how uncertainty affects scientific conclusions.

### Skills

A third element of change that has been documented regularly is an increase in skills related to the project. Examples include identification skills (Shirose et al. 1997, Knutson et al. 1999, Genet et al. 2008), experimental design skills (Trumbull et al. 2000), data-entry and analysis skills (Bonney 2007), and presentation skills. Additionally, involvement in citizen science can result in fostering ways of thinking that are consistent with those of professional scientists and that are crucial for decision making and for developing ecological literacy (Trumbull et al. 2000, Jordan et al. 2011).

### Attitudes

Less investigated are changes in individuals' attitudes toward the scientific process (though, as noted above, many participants already have high levels of interest in science and, presumably, a postive attitude toward it). However, involvement in citizen science has been found to increase appreciation of science (Wilderman et al. 2004, Price and Lee 2013) and of the role of scientists in contributing to policy. Participants may experience an increase in their self-esteem associated with their role as a volunteer (Rohs et al. 2002). Participants may also develop a greater sense of stewardship over the populations or sites they are responsible for surveying or monitoring (Carr 2004), and greater confidence in expressing their concerns about local environmental issues (e.g., Leslie et al. 2004, Nerbonne and Nelson 2004).

### Behaviors

Of particular interest to practitioners are changes in behaviors related to the specific focus of the project and to "nature" more generally. Changes in behaviors associated with political activism are also of interest but are difficult to determine given the longer-term follow-up needed and the specificity usually associated with purposeful behavior-change campaigns. Some behavioral changes have been reported in individuals who participated in a citizen science project, such as engaging in a greater number of conversations with others regarding the object or process being studied than they did prior to their participation (Overdest et al. 2004, Evans et al. 2005). Furthermore, individuals may increase the number of days on which they observe studied organisms and record data after participation in a project (Thompson and Bonney 2007). Participants also have exhibited increased awareness of their backyards as habitat and have changed aspects of their behavior in regard to their yard and its management (Evans et al. 2005). Finally, individuals and communities that have participated in citizen science projects have increased their efforts to influence policy regarding the object or process being studied (Overdest et al. 2004).

### Other Beneficial Changes

There are other changes associated with participation in citizen science projects, particularly when the project is focused on a community. These changes include increased social capital (Adger 2003, Overdest et al. 2004), network building (Leslie et al. 2004, Nerbonne and Nelson 2004, Overdest et al. 2004), increased community capacity (Conrad and Daoust 2008), positive economic impact (Gardiner et al. 2012), increased feelings of community connectedness (Overdest et al. 2004, Fernandez-Gimenez et al. 2008), and increased trust between scientists and land managers (Fernandez-Gimenez et al. 2008). Finally, both participants and nonparticipants may derive additional benefits from preservation and sustainability outcomes associated with citizen science projects (Wilderman et al. 2004, Evans et al. 2005, Pandya 2012).

## GAPS IN OUR KNOWLEDGE

Additional work is needed to examine who does and does not participate in citizen science projects and the associated opportunities and barriers. Past research has examined the inclusion of historically underrepresented groups in citizen science but has not considered upcoming demographic and social changes in the United States. These changes include the increasing percentage of children under the age of eighteen who have immigrant parents (Jolly 2002). Are new immigrants more or less likely to become involved in citizen science projects? Do we inadvertently exclude some of these individuals by labeling our programs as "citizen science" rather than "community science" or simply "public science" programs? If we are to be truly inclusive in terms of a diverse pool of participants, should there not be consideration of how welcoming our programs are to the LGBTQ community? There appear to be no data pertaining to participation in citizen science programs by LGBTQ individuals. Are there social barriers that are invisible to scientists but that may exist for members of various communities? Identification of such barriers is the first step in eliminating them.

Another knowledge gap that needs to be addressed is demonstration and acknowledgment of what scientists learn from participants in citizen science projects. One important exception is Leslie et al.'s (2004) study on using volunteers in fisheries projects, in which the authors state that scientists learned and continue to learn supervisory and communication skills from various volunteers.

## SUMMARY

Participation in citizen science can result in many beneficial changes to individuals and communities. These include increases in awareness of scientific issues that impact people at different scales and knowledge about the organism or process under study and the scientific process itself. Furthermore, individuals and communities may develop more positive attitudes toward science, scientists, and scientific thinking. Several skills may also be developed or enhanced, including identification skills, data-entry skills, analysis skills, and presentation skills. Finally, though not as widely documented to date, behavioral changes can also occur. These changes in behavior include increases in conversations about scientific issues, in observations and recording of study organisms or processes after a project ends, and in efforts to affect policy. Beyond the potential and observed changes identified here, there are many aspects of citizen science and its effects on participants to explore further.

# PART THREE

# CITIZEN SCIENCE IN PRACTICE

# FOURTEEN

# From Tiny Acorns Grow Mighty Oaks

## What We Have Learned from Nurturing Nature's Notebook

THERESA M. CRIMMINS, LORIANNE BARNETT, ELLEN G. DENNY,
ALYSSA H. ROSEMARTIN, SARA N. SCHAFFER, and JAKE F. WELTZIN

The timing of plant and animal life-cycle stages within a season, termed *phenology,* encompasses phenomena such as leaf-out and flowering, insect emergence, and bird migration. Phenology can be used as an indicator of species sensitivity to environmental variation and can also be combined with climatological data as an integrated indicator of climate change. Historical phenology datasets have been critical to documenting the impacts of climate change on biological systems at both national and global scales, and as such, phenology has been called the "fingerprint of climate change impacts" (Parmesan 2007).

Because of the tremendous importance of phenology data and information, the USA National Phenology Network (USA-NPN; www.usanpn.org) was established in 2007 by the U.S. Geological Survey (USGS) in collaboration with other governmental and nongovernmental organizations as a national-scale science and monitoring initiative (Schwartz et al. 2012). Stakeholders of USA-NPN include researchers, resource managers, educators, communication specialists, nonprofit organizations, human health organizations, science networks, and the public. The primary goals of USA-NPN are (1) to improve understanding of how plants, animals, and landscapes respond to environmental variation and change; and (2) to inform decisions pertaining to land use, risk management, and natural resource management (USA-NPN National Coordinating Office 2014). The efforts of USA-NPN are organized and directed by the staff of its National Coordinating Office (NCO).

Phenology datasets that are best suited for supporting scientific discovery and decision making are those that consist of observations of multiple life-cycle stages collected at regular intervals at the same locations over multiple years. The NCO collects, stores, and shares high-quality observations of plant and animal phenology at a national scale by engaging observers in Nature's Notebook (figure 14.1; www.naturesnotebook.org), a national-scale,

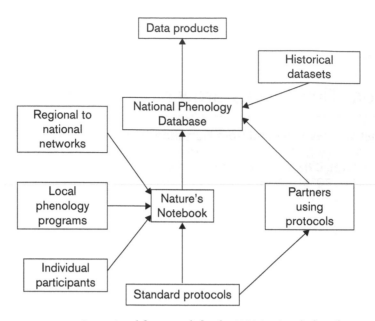

FIGURE 14.1. Operational framework for the USA National Phenology Network (USA-NPN). The foundation of USA-NPN's online phenology observing system, Nature's Notebook, is standardized protocols for monitoring phenological activity of plants and animals. Participants in Nature's Notebook include those observing phenology independently, those contributing to a Local Phenology Program recognized by USA-NPN, and those contributing to another regional or national network. Data collected through Nature's Notebook are stored in the National Phenology Database and are used to support the production and delivery of data products. Other organizations can use the standard protocols with their own observing systems, and their data can be shared directly with the database. Historical data can also be contributed to the database.

multi-taxon phenology observing program appropriate for both professional and volunteer participants. Participants in Nature's Notebook report on the phenological status of individual plants and of animal species at a location, following rigorous and standardized protocols, and are encouraged to make regular observations over the course of multiple years. Observations are stored in a national database and are used to produce value-added products such as maps, calendars, and reports (USA-NPN National Coordinating Office 2016). Data are used for a variety of applications related to science, conservation, and resource management, including predicting the spread of invasive species, validating information derived from remotely sensed imagery, and disentangling the environmental drivers of leaf-out.

With ten years of experience collecting, storing, and sharing a high-quality, national-scale data resource of plant and animal phenology status observations, we have faced a variety of challenges, including (1) recruiting and retaining Nature's Notebook participants;

TABLE 14.1 Challenges and Mitigation Strategies for the USA National Phenology Network

| Challenge | Mitigation |
|---|---|
| Participant recruitment | Place emphasis on engaging three primary audiences, including individuals and groups or institutions. Carefully craft messages with a focus on how potential users can meet their own goals through monitoring phenology. |
| Participant retention | Communicate frequently with participants. Express gratitude frequently. Demonstrate how data are being used. Provide meaningful engagement materials. Facilitate a community of practice for phenology leaders. |
| Data quantity | Create a national-scale observing program that engages a diversity of contributors. Host campaigns that focus on specific species. |
| Data quality | Provide standardized protocols with detailed descriptions. Provide a variety of training materials, with an emphasis on consistency in data collection. Implement methods to prevent transcription errors. Perform latitude/longitude checks of observation locations. Flag conflicting records. |
| Operational sustainability | Operate under a organization-wide, long-term strategic plan. Operate under a National Coordinating Office–specific, short-term implementation plan. Maintain a risk management plan. Produce annual reports to demonstrate successes. Maintain terms of use. Ensure that data are used in science and management applications. |

(2) ensuring that our long-term observations are of the highest possible quality; and (3) growing and sustaining the program (table 14.1). In this chapter, we share some of the insights that made the greatest difference to our success over the years. The lessons shared herein are likely to be most relevant to citizen science projects that (1) engage volunteers in collecting field-based biodiversity observations, (2) operate at the regional or national level, and (3) seek data of sufficient quality to support scientific research, especially where frequent and repeated observations over time are of greatest value.

## BUILDING A LEGION OF OBSERVERS: HOW DO WE GET THE DATA WE NEED?

The most valuable data for documenting how species and systems are responding to variable and changing climate conditions consist of regular, repeated, careful observations of

FIGURE 14.2. Leaf emergence, leaf elongation, and fruit growth in fig (*Ficus* sp.). Credit: photos by Sara N. Schaffer.

multiple life-cycle stages at the same locations over time (figure 14.2). Nature's Notebook participants are encouraged to make and submit observations of one or more plants and/or animals at their monitoring site approximately once a week, during the portion of the year when the species under observation are active (figure 14.3). Ideally, participants collect observations over the course of multiple years. The most useful data are those consisting of frequent observations of the same individual plants or the same species of animals at the same location. Further, the broadest set of science and management questions can be answered with dense geographic coverage of observations across the United States for many species.

Recognizing that we are asking quite a lot from our participants—frequent observations spanning multiple years—we seek observers with a high level of skill and commitment. Our goal is to engage three primary audiences (as shown on the left side of figure 14.1): (1) citizen scientists who participate independently in Nature's Notebook; (2) leaders and members of Local Phenology Programs (LPPs); and (3) staff or affiliates of established, formal, regional-to national-scale observing networks such as the National Park Service Inventory and Monitoring Program. For all three audiences, our recruitment tactics are to be welcoming and honest about the level of commitment needed. Our retention tactics include providing frequent, genuine appreciation and clearly demonstrating how participants' contributions are being used through communication channels that we can manage at a national scale. The approaches we use to sustain participation among independent participants also support individuals participating as part of LPPs and as part of formal regional or national observing networks. We offer additional support for leaders of LPPs and of formal observation net-

FIGURE 14.3. A Nature's Notebook participant records an observation of wildflower phenology. Credit: photo by Brian F. Powell.

works, including a community of practice, opportunities for social interaction, and enhanced training and engagement materials, including program planning and leadership development courses.

### Recruiting Participants

As we endeavored to build a cadre of observers to generate the spatial and temporal density of observations that we seek, we learned the importance of thoughtfully identifying our target audiences and approaching them with customized messaging focused on how they might benefit by observing phenology. We have realized that spreading an invitation to join Nature's Notebook without a targeted audience is a waste of resources and that advertising our program broadly to the general public yields few committed observers (box 14.1). Reflecting on these experiences, we realized the critical importance of clearly identifying and articulating to potential participants what they stand to gain by tracking phenology. We have since adopted an approach to recruitment that involves identifying individuals or groups of people that have the greatest potential for being committed to tracking phenology and then approaching them with their needs in mind. When reaching out to a new group of potential participants, we now emphasize how phenology information might help them accomplish their own goals, including better understanding backyard gardens or habitats, enhancing community outreach and education, and supporting local decision making or research.

### Retaining Participants

Once we recruit active participants to Nature's Notebook, we do not want to lose them! These individuals are our greatest opportunity for growing our data resource. Not only does

We discovered the importance of identifying—or segmenting—our audiences, which is critical for effective resource allocation to recruitment campaigns, somewhat serendipitously. We were thrilled when the American Association of Retired Persons (AARP) contacted us in 2012 and indicated they would like to include a blurb on our phenology monitoring program in their glossy circular, distributed to eleven million adults. We were excited, anticipating a huge surge in activity in our program—so much so that we created a prominent "Welcome, AARP!" banner on our homepage, with a link to a special landing page for members of the association. However, over the four months following the distribution of the circular, our AARP page was visited by fewer than 250 unique individuals, or about 0.1% of our total website traffic over the period, resulting in just fifty-four Nature's Notebook recruits.

By contrast, on the basis of personal relationships established with the leaders of the Community Collaborative Rain, Hail, and Snow Network (CoCoRaHS), a citizen science program, we were invited to include an invitation to join Nature's Notebook in their Message of the Day—sent to their volunteers, who were otherwise focused on observing precipitation. This invitation resulted in hundreds of recruits to Nature's Notebook. Moreover, it turns out that CoCoRaHS observers who participate in Nature's Notebook are among the most consistent participants in our program. We believe this is because they already participate in a similarly rigorous citizen science project aimed at generating long-term data. CoCoRaHS observers appreciate the importance of following established protocols, making repeated observations at their location, and ensuring that these observations are submitted online in a timely fashion, and they apply the same rigor to phenology monitoring.

long-term participation yield longer-term, site-specific observation records—and therefore more valuable data for documenting trends or changes in the timing of plant and animal phenology—but retaining participants from year to year reduces the amount of time needed to find and train new participants. Further, the greater experience boasted by long-term participants may lead to higher accuracy in the data provided by these individuals. Finally, long-term participation may lead to greater satisfaction for participants.

Social science research stresses the importance of offering authentic, challenging, and meaningful activities and engagement materials for keeping participants active and committed to a program. These materials should engender experiences that center on problem solving. Recent research into problem-solving experiences (or project-based learning) has demonstrated that those who engage in this type of learning gain and retain information more effectively (Darling-Hammond et al. 2008). Bearing this in mind, we designed Nature's Notebook outreach materials to help participants find personal meaning in their phenology observing experience.

Given that we are a national-scale program with a small support staff, we are limited in our ability to provide one-on-one communication with individual observers. To maintain a connection with our observers, we use a variety of Internet-based approaches and tools, including newsletters and email messages, social media, listservs, periodic webinars, and regular updates to our website. Each of these forms of communication is available to all participants in Nature's Notebook, regardless of whether they participate as an individual or as part of an established group.

Because we rely heavily on digital forms of communication, it is critical that these communications be interesting and information-rich. We have committed to being genuine, generous, and grateful in our messaging (Miller 2013), and we occasionally seek feedback from participants via short surveys to further determine their interest. We try to keep participant needs and interests in mind at all times. For example, our newsletters incorporate news tidbits that we wish to convey and also focus on topics most likely to be of interest to participants based on their feedback. We also provide timely results of participant contributions and interpretation appropriate for nonscientists (see examples at www.usanpn.org /newsletters). Finally, we translate peer-reviewed publications into language that is accessible to participants, posting to our website clear and concise summaries of publications that used data provided by Nature's Notebook participants (see www.usanpn.org/nn/vignettes and www.usanpn.org/nn/connect/highlighted_pubs).

In the interest of maintaining enthusiasm and activity among participants and offering multiple ways to participate, we offer featured data-collection "campaigns" focused on particular species or topical applications, with the goal of accumulating sufficient observations to answer specific research or management questions (www.usanpn.org/nn/campaigns). Often initiated by a researcher with a specific data need, these campaigns typically run for three to five years. We strongly encourage researchers to work with us to communicate directly with Nature's Notebook participants via webinars and to provide data summaries and updates through periodic campaign reports. The recent "PopClock" campaign (www .usanpn.org/nn/popclock), conducted in partnership with the University of Maryland Center for Environmental Science (UMCES), serves as an excellent example of this model. Researchers at UMCES engaged Nature's Notebook observers in collecting observations of leaf phenology for two species of poplar (*Populus* spp.) to support global change research and published the results of their study in a peer-reviewed journal (Elmore et al. 2016).

Though we are still compiling data to determine the efficacy of the campaigns and other retention tactics, a recent analysis suggested that frequent, information-rich newsletters delivered via email maintained higher activity levels among Nature's Notebook participants, at least in the short term (Crimmins et al. 2014). As such, we will continue with these approaches, and undertake periodic evaluation of their impacts to inform our activities related to recruiting and retaining participants.

## Models of Group Engagement

In addition to engaging individual participants in Nature's Notebook, we encourage participation among members of established volunteer groups, not-for-profit organizations, public research and land management agencies, and other research or monitoring networks. These various groups fall into two broad categories: LPPs and regional or national observing networks. A major strength of these forms of participation is that because phenological

FIGURE 14.4. Members of several Local Phenology Programs convene at an in-person training workshop. Credit: photo by LoriAnne Barnett.

monitoring is shared by multiple individuals, the duration of the activity has the potential to extend beyond the tenure of any one individual.

### Local Phenology Programs

A rapidly growing segment of Nature's Notebook observers participate as part of LPPs or other locally to regionally organized groups focused on, or incorporating, phenological monitoring (figure 14.4). Many organizations, including nature centers, arboreta, schools and colleges, Master Gardener and Master Naturalist chapters, and land conservancies and trusts have recognized Nature's Notebook as an interesting and relevant program to build into their science, education, and outreach programming. In 2018, two hundred LPPs submitted observations of phenology to Nature's Notebook. We encourage such groups or institutions to identify a locally relevant, tractable research or management question that their phenology observations could help answer, while contributing to our national database. This approach enables participants to understand how their site-specific data are being used or applied at a range of scales. In addition, we assume that when phenological monitoring is used to directly support the science, management, or outreach goals of an organization—as

FIGURE 14.5. Participants in a Local Phenology Program work together with a mobile application, observing phenology. Credit: photo by LoriAnne Barnett.

opposed to being used simply as an ephemeral activity—there is a greater likelihood that the monitoring activities will be sustained.

In terms of data quality and quantity, LPPs offer an important approach for helping us grow our phenology database (figure 14.5). Because data collection is typically distributed across multiple observers, the burden to collect and submit data is less for members of LPPs than for individuals participating in Nature's Notebook independently. Further, participants have the support of other group members, which may encourage continued participation. Indeed, individuals who participate in Nature's Notebook as part of an LPP are likely to submit more total observations, remain active in the program beyond their first year, and submit higher-quality data than those who participate in Nature's Notebook independently (USA-NPN NCO, unpublished data). We believe that LPPs produce more and higher-quality data because they typically offer face-to-face training, support, opportunities for camaraderie, and social interaction. LPPs may provide other benefits to participants, including increased understanding of the natural world and the scientific process, as well as opportunities to interact directly with scientists and research staff at the organization that hosts the LPP.

Because LPPs produce comparatively large amounts of high-quality data, it is imperative that we do what we can to ensure they remain active. Thus, we nurture a *community of practice* for leaders of LPPs. Communities of practice are groups of people who share a common interest and who experience increased expertise around the topic as a function of interacting with the group (Wenger 1998). Many potential LPP leaders—whom we recognize as Local Phenology Leaders (LPLs)—ask us how others have succeeded in establishing and maintaining a phenology program. We encourage leaders to share their lessons learned with others in the field through the LPL email listserv. Through this listserv, LPLs can interact, share ideas and lessons learned, and offer support to others. To help grow and sustain the LPL community of practice, we offer periodic webinars tailored specifically to LPLs, have dedicated a portion of each bimonthly newsletter for network partners *(The Connection)* to LPLs, and offer an intensive ten-week (sixty- to eighty-hour) LPL Certification (www.usanpn.org/nn/LPLCertification).

A major challenge we face is the inevitable turnover in LPLs. When enthusiastic program advocates leave an institution that is hosting an LPP, this can leave a vacuum in leadership and groups can falter. To mitigate this risk, we encourage LPLs to ensure that the program directly and explicitly accomplishes its organizational education, outreach, or management goals through program planning, thereby tying the activity of the LPP to the mission of the host institution. If the value of the program can be seen readily by other staff members, this may improve the likelihood that the LPP will be sustained.

### Regional or National Observing Networks

A third approach we take to building the phenology database is to encourage the inclusion of phenology monitoring as part of existing long-term natural resource monitoring, or in support of research activities conducted by institutions, agencies, and their affiliates (figure 14.1). Many of these institutions are regional or national in scale, including the National Ecological Observation Network (NEON; Elmendorf et al. 2016), sites within the Long-Term Ecological Research Network (LTER), U.S. Fish and Wildlife Service National Wildlife Refuges, and units within the National Park Service (e.g., Tierney et al. 2013). A key feature in each case is that the institution decided to conduct phenological monitoring to meet their own management or research objectives (figure 14.6).

USA-NPN provides a hierarchical, national monitoring framework that enables other organizations to leverage the capacity of the network for their own applications, thereby minimizing investment and duplication of effort, while promoting interoperability of data. Network participants can leverage several tools and services offered by USA-NPN, including standardized monitoring protocols, long-term data maintenance and archiving via the

FIGURE 14.6. A refuge manager observes the phenological status of a cottonwood tree at the Valle de Oro National Wildlife Refuge in Albuquerque, New Mexico. Credit: photo by Erin Posthumus.

National Phenology Database, and all of the outreach, education, and training materials created for Nature's Notebook. In some cases, we have also worked with institutions to develop organization-specific sections on the USA-NPN website (e.g., www.usanpn.org/fws).

## HOW DO WE ENSURE THE QUALITY OF OUR DATA?

### Encourage the Use of Standardized Protocols

Until recently, observations of phenology in the United States were collected through independent efforts using a variety of definitions and methods, making comparative analyses

Milkweed
*Do you see...?*
- Young leaves
- Leaves
- Colored leaves
- Falling leaves
- Flowers or flower buds
- Open flowers
- Fruits
- Ripe fruits
- Recent fruit or seed drop

Monarch
*Do you see...?*
- Active adults
- Flower visitation
- Migrating adults
- Mating
- Active caterpillars
- Caterpillars feeding
- Dead adults
- Dead caterpillars

FIGURE 14.7. An example of phenology observations collected on a monarch and a pineneedle milkweed through Nature's Notebook. Credit: photo by LoriAnne Barnett.

across studies and species challenging. Geographically extensive observations, collected using standardized protocols, offer a stronger resource for advancing science and informing management decisions (Schwartz et al. 2012). To improve coordination in phenology data collection at a national scale, we engaged scientists, resource managers, and educators to develop standardized phenology observation protocols for plant and animal species across the diversity of ecological systems (Denny et al. 2014). These protocols are rigorous enough to support scientific discovery, general enough to be applicable to many different species, and straightforward enough to be used by professional and citizen scientists alike (figure 14.7). The protocols also enable other organizations to collect phenological data using the definitions in the standardized protocols, but using their own interface, whereupon they can be shared and interoperable with the National Phenology Database (e.g., Elmendorf et al. 2016). The protocols were also designed to facilitate integration with historical phenology datasets collected according to different protocols, although integration can be difficult and

FIGURE 14.8. Phenophase progression for a citrus species. Credit: photos by Patricia Guertin.

time-consuming, depending on how different the historical protocols are from the contemporary protocols (Taylor et al. 2019).

### Ensure and Control Quality

We define *quality assurance* as approaches used to minimize the likelihood that data are incorrectly collected or entered into the database, and *quality control* as approaches used to determine the validity of the data once it has been entered into the database (USA-NPN National Coordinating Office 2016).

#### Quality Assurance

Approaches we use for quality assurance focus on improving species identification, providing clear information and training about how to evaluate the phenological status of an organism, and minimizing errors in the process of data entry. Even with clear, standardized observation methods, it can still be difficult to ensure consistency in phenology observations because evaluating phenological status can be challenging. First, observers must be able to accurately identify the species under observation. Second, to assess phenological status, observers must be familiar with the botanical structures (buds, leaves, flowers, fruits) or animal stages or behaviors for the species they are observing. Further, pinpointing the date at which a plant transitions from a breaking leaf bud to a leaf or from a fresh flower to a wilted flower is somewhat subjective (figure 14.8). For example, different observers can make different decisions about the exact date when a fruit transitions from unripe to ripe, leading to inconsistencies across observers.

One way to evaluate how well Nature's Notebook participants perform in consistently identifying phenological stages is to compare their reports to those collected by a designated expert at the same time. A recent formal comparison revealed that Nature's Notebook observers can identify plant phenophase status with greater than 90 percent consistency and can pinpoint phenophase transitions with approximately 70 percent consistency (Fuccillo et al. 2015).

To minimize inconsistencies among observers, we have designed and incorporated several approaches into the observation methods and training materials. First, we provide carefully worded definitions for each phenophase (Denny et al. 2014). We stress that observers should interpret the definitions literally, and we encourage observers to report only on the phenological stages they are comfortable identifying. Second, we offer detailed descriptions and photographs of phenological stages in a wide variety of training and observer support materials (e.g., Guertin et al. 2015). We make these materials readily available on our website and available for download as documents, videos, and webinars for use by individual observers and LPLs. Other approaches have been built into the Nature's Notebook user interfaces to reduce errors in species identification, data entry, and transcription (Rosemartin et al. 2018a).

### Quality Control

To ensure that incoming data are as error-free as possible before they are shared with potential users, we undertake several automated quality control checks on our data. These checks include (1) verifying that latitude and longitude coordinates fall within North America, (2) validating elevation on the basis of coordinates, and (3) flagging conflicting phenology status observations for the organism on the same day. We are currently working to flag observations where phenological stages occur in an implausible order.

We have considered many additional, logical quality control tests for our records, for example flagging identifications outside of a species' published range or phenophases occurring earlier or later than expected. However, for many of these tests, we do not yet have the quantity of data to enable the identification of obvious errors or reports that are significantly out of range. More importantly, outliers may not always represent inaccuracies. Rather than signaling errors in the data, the unusually early appearance of an open flower or leaf, or the report of a bird species farther north than expected, might represent the first signs of changes in phenology or species distribution. Other approaches have been built into the National Phenology Database to maximize data quality, with a focus on validation, reliability, and plausibility. We track data quality via an annual quality-control report and dashboard (Rosemartin et al. 2018a; www.usanpn.org/data/quality).

## PROGRAMMATIC CHALLENGES

### Meeting the Needs of a Diversity of Stakeholders

USA-NPN was originally envisioned and established as a broad network composed of partnerships between governmental and non-governmental science and resource management agencies and organizations, the academic community, the public, and other stakeholders.

Motivations for participation were expected to be broad, ranging from research to education and outreach, to applications related to agriculture, tourism and recreation, human health, and natural resource conservation and management. To build network participation and buy-in, we held a number of planning and implementation workshops, where we focused strongly on understanding the motivations, roles, and responsibilities of potential stakeholders and network participants (Betancourt et al. 2005, 2007; Nolan and Weltzin 2011; Enquist et al. 2012). Ultimately, to acquire the data needed to meet our goals with a relatively small budget and staff, we found it necessary to be a big-tent organization with room for many types of stakeholders and affiliated projects.

USA-NPN engages in boundary work, bringing together scientists and nonscientists as well as disciplines that do not typically work together, which requires careful consideration of customs and language that differ among disciplines (Guston 2001, Cash et al. 2002). Over the past ten years, the network has grown to include a broad coalition of stakeholders, from teachers using Nature's Notebook to teach about organisms' responses to climate change, to academic and federal researchers in fields as diverse as ecology, biogeochemistry, and climate science, to park staff engaging the public in research on invasive species.

The diversity of stakeholders represents a continuing challenge because each member has different motivations for participating. To meet this challenge, we have focused on identifying and providing common services, enabling us to scale up effectively and meet many needs at once. However, because of the diversity of stakeholder needs, we have found it necessary to carefully segment our audiences and to develop content and marketing specific to each. For example, some resource management organizations charged with monitoring are interested in using the Nature's Notebook platform for data collection and management. By contrast, other resource management organizations are most interested in our value-added data products, such as phenological forecasts, maps, and calendars.

At the initiation of the NCO in 2007, the small staff consisted almost entirely of formally trained scientists. Although this composition facilitated communication with the scientific community and, to a lesser extent, with the resource management community, it soon became apparent that different skill sets were required to develop tools and services, to manage and add value to data, to market the program, and to recruit and retain thousands of volunteer participants across the nation. We addressed these needs through a combination of hiring and contracting, by cross-training staff and by developing external collaborations that contributed novel expertise and resources. Today, our staff represents a balance of scientists, educators, program developers, and practitioners, and our program benefits from a diversity of backgrounds and approaches to problem solving.

### Planning, Implementing, and Reviewing the Program

Since 2007, USA-NPN and the services and data provided by the NCO have grown quickly. Early guiding documents included a number of broad strategic and implementation planning documents (Betancourt et al. 2005, 2007). Over the years, our planning framework has evolved to include a strategic plan, which provides the long-term (thirty-plus-years) vision for the network; an implementation plan, which guides the NCO staff on shorter-term (one- to five-year) actions; a risk management plan, which forces staff to consider contingencies for potential threats to the organization; annual reports, which summarize our accomplishments for the year at a level appropriate for a wide audience; and a data product catalog, which outlines a five-year plan for development and delivery of data products for different stakeholders.

Our strategic plan articulates the long-term potential of the network to serve scientists, resource managers, policymakers, and the U.S. public (USA-NPN National Coordinating Office 2014). Although this plan took considerable time and effort to produce, it is an important tool for bringing the various components of the network (i.e., staff, advisors, funders, existing partners) together to agree upon a shared long-term vision and mission. The strategic plan lays out an ambitious framework for understanding how phenology is changing, the causes and consequences of change, and what these changes mean for natural resource managers. We have found that this broad, grand vision inspires our partners to contribute their effort, intellect, and financial resources to the cause.

Our implementation plan focuses staff effort on measurable, achievable outcomes that move the broader network toward our overarching goals. Periodic assessments of our progress using these measures enables us to evaluate our efficacy and helps us identify mistakes, misconceptions, and failures in our approach. Another realization has been that while the implementation plan has been very effective for guiding staff efforts and ongoing evaluation following the logic model (Kellogg Foundation 2004), the strategic plan has been more effective in inspiring partners and garnering funding.

Finally, we subject our program to periodic formal programmatic reviews. Our first such review took place in 2014 (Glynn and Owens 2015) and led to recommendations that changed the course of USA-NPN (USA-NPN National Coordinating Office 2016). This form of feedback has had great value for shaping our foci and activities. We intend to undergo formal external reviews every five to ten years.

### Identifying and Managing Liability and Risk

Key approaches to managing risk include identifying, describing, and limiting—where possible—liability. Because Nature's Notebook engages participants who are distributed across the nation, with limited oversight by the NCO, it was critical to establish comprehensive website terms of use (www.usanpn.org/terms). Broadly, the terms of use describe roles and responsibilities, and limitations thereof, for all people and organizations that interact with the program, for example how content is owned, managed, and shared; how participants may interact with the program via web-based and mobile applications; and how privacy is maintained. The terms of use also include policies related to data use and data attribution, and policies governing use of Nature's Notebook or its components by participants and third parties.

Identifying and evaluating a wide range of risks—from loss of funding to loss of key personnel to competition from other organizations—is also an important component of the organizational planning process. Once the impact, likelihood, detectability, and preventability of major risks have been assessed, risks can be prioritized, delegated to responsible indi-

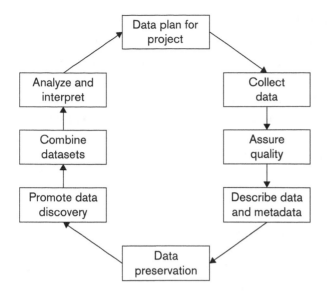

FIGURE 14.9. The data life cycle. Credit: adapted from Wiggins et al. (2013).

viduals, and assigned to one of several potential outcomes: accepted, transferred, avoided, or mitigated. To meet this need, we developed an internal risk management plan, which plays a key role in communicating and managing the vulnerabilities of the NCO among staff, advisors, and funders. Though risk management plans should be reexamined and modified regularly, they are part of an effective strategy for delineating, prioritizing, and mitigating potential issues to ensure continuity of operations and long-term sustainability.

### Creating Sustainable Data: Data Are Forever

Chartered to develop a national-scale, long-term data resource, USA-NPN must consider sustainability in terms of data documentation, accessibility, and traceability. We use the framework of the data life cycle to guide both short- and long-term operations related to data and information. The data life cycle (figure 14.9) encompasses all facets of data generation's path to knowledge creation, including planning, collection and organization of data, quality assurance and quality control, metadata creation, preservation, discovery, integration, and analysis and visualization (Michener and Jones 2012, Wiggins et al. 2013). As a federally funded effort, we have ensured our compliance with federal data policy throughout the data life cycle (Rosemartin et al. 2018b). To ensure the availability of the data that we curate beyond the life span of USA-NPN, we share data and metadata with several community repositories, including the Knowledge Network for Biocomplexity, DataONE, the USGS Science Data Catalog, and the USGS data repository ScienceBase. Well-documented and widely available data-management resources such as these enable science on heretofore unknown spatial and temporal scales (Hampton et al. 2013, Peters et al. 2014, Soranno and Schimel 2014) and help ensure the longevity and sustainability of data for future generations.

### Establishing Scientific Credibility

Strong science, robust data, sustained funding, and a secure infrastructure do not ensure a program's success. Credibility is also important. In the science and management communities, the term *citizen science* can connote data collected haphazardly, with little rigor or training (Boudreau and Yan 2004). In truth, citizen science can contribute substantially to scientific research, conservation, and resource-management decision making (Dickinson et al. 2010, Theobald et al. 2015, McKinley et al. 2017).

The data resources assembled by USA-NPN are collected by both citizen scientists and professionals at federal agencies and universities; about one-third of the data in the database were collected by professionals (USA-NPN NCO, unpublished data). To ensure the credibility of all data stored and shared through USA-NPN, the staff of the NCO worked closely with the first few contributing researchers to analyze the data and minimize misinterpretation of data. Between 2007 and 2018, sixty papers using contemporary or historical data collected and managed by the NCO were published in peer-reviewed journals, ameliorating many concerns about the perceived quality and usefulness of the data (www.usanpn.org/pubs /results).

Phenology data managed by USA-NPN have been used in a wide range of science applications. For example, a historical lilac and honeysuckle dataset we house (Rosemartin et al. 2015) was used to show that 2012 was the earliest spring over the continental United States in the past 112 years (Ault et al. 2013). Although the cumulative effects of this unusually early spring were most pronounced across the Corn Belt, the western Great Lakes region, and the northeastern United States, the beneficial effects of an early spring were subsequently offset by a late spring frost and summer drought. Likewise, data we host were used to develop phenological models that showed that the length of the growing season over the past thirty years has increased by about 3–4 percent in the eastern deciduous forest, New England, and northern Rocky Mountain forests (Yue et al. 2015). Many other examples are described on the USA-NPN website (www.usanpn.org/nn/vignettes).

### Seeking and Maintaining Funding

The NCO was initially funded, either directly or indirectly (e.g., through in-kind contributions such as staff), by a number of sponsoring organizations and agencies. For example, the National Science Foundation provided resources through a Research Coordination Network grant that enabled a number of workshops that set the stage for planning and implementation of the network. Since that time, major funding has been provided by USGS, with contributions from other agencies in the Department of Interior (particularly the U.S. Fish and Wildlife Service and the National Park Service) as well as NASA. As such, it is critical that USA-NPN—although envisioned by a broad coalition of stakeholders—demonstrates value to the missions of the Department of Interior and USGS. Notably, however, this necessary emphasis on the priorities of major funders is not inconsistent with the missions of other stakeholder organizations. Overall, the clear vision and mission statements set in our strategic plan have helped us stay focused on where we should and should not apply our efforts for maximum benefit to our large group of stakeholders, enabling us to prioritize activities in the face of many potential opportunities that might otherwise have been distracting.

## SUMMARY

In the process of establishing a national organization running a citizen science program, we have learned many lessons specific to engaging individual and organizational participants, developing high-quality data resources, managing risk, sustaining our infrastructure, engaging data users, and more. We have come far in a relatively short time, but we realize there are many more lessons waiting to be learned as our organization, and the field of citizen science more broadly, continues to mature. We hope others will learn from our experiences, and we look forward to continued engagement with the citizen science community in the future.

## ACKNOWLEDGMENTS

We are thankful to the many participants who contribute to Nature's Notebook. The project described in this publication was supported by Cooperative Agreement nos. G14AC00405 and G18AC00135 from the U.S. Geological Survey. Any use of trade, firm, or product names is for descriptive purposes only and does not imply endorsement by the U.S. government.

# FIFTEEN

# Citizen Science at the Urban Ecology Center

## A Neighborhood Collaboration

JENNIFER CALLAGHAN, BETH FETTERLEY HELLER, ANNE REIS-BOYLE,
JESSICA L. ORLANDO, and TIMOTHY L. V. VARGO

### THE BEGINNING

Setting up a citizen science program was not part of the initial vision for the Urban Ecology Center (hereafter "the Center"), so you might say we stumbled into the whole field. The Center grew out of a neighborhood effort to fight crime and revitalize a neglected green space and was focused on neighborhood-based environmental education. In 2000, a part-time employee approached management with the concept of developing a vibrant urban research program in Milwaukee. She secured short-term grant funding to set up research projects guided by professional scientists, which engaged the public.

Initially, the goal of the program was to connect community members, decision makers, voters, and citizens with what was happening in academia. If there was cutting-edge ecological research, we wanted to bring it to the general public. Further, we decided early on that we were more likely to be successful if universities were involved, so we created a professional advisory committee with representatives from the University of Wisconsin–Milwaukee, the University of Wisconsin–Madison's Arboretum, Cornell University, and other institutions. The committee met twice a year to advise the research.

### THE MIDDLE YEARS: PROGRAM EXPANSION

What had started as a short-term, grant-funded program soon became a permanent program included in the Center's operating budget, consisting of a full-time manager working with student interns and over a hundred community scientists. Initial efforts were focused on increasing the number of citizen-based monitoring programs the Center could offer—

efforts that were strongly supported by the Wisconsin Department of Natural Resources' Citizen-based Monitoring Partnership Program. Initial programs used protocols and equipment from established national programs, such as Monitoring Avian Productivity and Survivorship, the Monarch Larva Monitoring Project, and Cornell University's Celebrate Urban Birds. Soon we established our own protocols based on our highly visible, urban setting with high concentrations of hikers, bikers, dogs, and other critters. For example, when a high proportion of our Sherman mammal traps were being destroyed by raccoons looking for a tasty morsel, we designed protective covers made out of plastic containers, along with a paired analysis that determined the device did not affect capture rate.

As the Center grew to include two additional neighborhood branches, the citizen science program expanded to become a network of three urban field stations in which all research and monitoring is accessible to and advised by community scientists. The monitoring projects create a strong framework around which community scientists can conduct cutting-edge research, from studying metabolites in migrating birds to locating snake hibernacula with surgically implanted diodes.

## A GUIDING PRINCIPLE

A core component of the Center's educational methods for children is to have well-informed, ecologically literate parents, grandparents, teachers, and friends serve as mentors to urban youth. This core component comes out of Significant Life Experience Research (SLER), which asks what happens in a person's life that helps them develop an environmental awareness or ethic, whereby a person not only understands environmental issues, but behaves in a way that respects the environment. The body of SLER research finds that an individual who grows up connected to nature (e.g., running in a park, looking at flowers, climbing trees) is likely to develop environmental awareness (Chawla 1998). This is especially true if the person's experiences in nature are combined with being in a mentoring relationship with someone concerned about the environment. Our school programs help foster environmental awareness in children, as our citizen science programs are focused on lifelong learning and building mentoring relationships between adults and children.

## THE PRESENT: INCREASING THE BREADTH AND DEPTH
## OF COMMUNITY RESEARCH

Early opportunities for citizen science volunteers at the Center relied mostly on data collection, as is true for many citizen science programs. While data collection is important and often fun, it represents only one part of the scientific process. We had very few opportunities for volunteers to engage in the front end (hypothesis generation, advising research) or the back end (data management, data visualization, and dissemination of results) of the research cycle, which sparked a question: Were we working with *citizen scientists* or *citizen field technicians*? For example, if someone worked with us to clean a river, they would be able to see the results of their efforts that same day. But what about someone who had been banding birds with us for ten years—what would they have to show for their efforts?

Funding from the U.S. Environmental Protection Agency's Great Lakes Restoration Initiative allowed the Center to hire two new positions, a research coordinator and a data specialist. Both of these became permanent, full-time positions, enabling a concerted effort to expand the depth and breadth of experiences for community scientists.

Initially we invited community members to serve on the professional advisory committee as a way to expand their roles. However, we soon found that most community members felt too intimated sitting next to the "experts" to be able to contribute in a meaningful way. Now we have two separate committees that inform each other while advising our research. Having two committees is a more complicated process, but we believe this better captures a wide range of experiences and allows for a more comfortable space to which community members can contribute equally.

## BEGINNING WITH THE END IN MIND: DATA STEWARDSHIP IS PARAMOUNT

When facilitating the design of original research projects by students in the University of Minnesota's Driven to Discover program (chapter 16), we stress the importance of beginning with the end in mind. What data will you need to answer your question and how will they be collected? How will data be analyzed and visualized to determine if they support your hypothesis? If we were to create a brand-new citizen science program from scratch, we would similarly focus on data analysis first, so that we would be creating projects with the end goals in mind. The addition of a full-time data specialist allowed us to better steward the data, provide opportunities for community scientists to manage and visualize the data, and communicate the results and stories associated with understanding the land of our restored urban green spaces.

### Data Collection

The majority of our data are collected with paper data forms. Data are then entered electronically into various in-house and online databases specific to each project and scanned and stored for later reference. While we have had the ability to set up a field collection form on a handheld GPS (Topcon GMS-2 Pro) in the past, interest in using this device was low because it is not very user-friendly, particularly when used infrequently. In recent years, we have started transitioning to using smartphone applications for data collection in the field using geographic information systems (GIS). For some projects, we use the ESRI Collector for ArcGIS application to collect data in the field that are directly linked to an online geodatabase (ArcGIS Online) and can also be connected to desktop mapping software (ArcGIS).

### Data Management

The Center's Research and Citizen Science Program maintains multiple monitoring projects (e.g., focused on frogs, mammals, turtles, park use, and vegetation) for which data are stored in-house using Microsoft Access databases. Other projects are stored in online databases, many of which are used in local, national, and international research by project partners and include some level of user-defined data exploration and visualization. Additionally, these online databases allow us to access our data records and export them for further analysis. For example, acoustic data from bat surveys are submitted to the Wisconsin Bat Program, and dragonfly and damselfly data are submitted to the Wisconsin Odonata Society, both of which are maintained by the Aquatic and Terrestrial Resources Inventory (http://wiatri.net), a program within the Wisconsin Department of Natural Resources. All of the Center's weekly bird-walk data are submitted to the Cornell Lab of Ornithology's eBird project (https://ebird .org), and bird banding data are submitted to the U.S. Geological Survey's Bandit database (https://www.usgs.gov/software/bandit-software). For many years, we have managed monarch butterfly data on the Monarch Watch (University of Kansas, https://monarchwatch.org) and Monarch Larva Monitoring Project (University of Minnesota, https://monarchlab.org) databases. In addition to formal surveys, incidental observations of vegetation and wildlife by the public and staff can be submitted to the Center's iNaturalist account (www.inaturalist .org/projects/urban-ecology-center-phenology), a project that has grown with the advancement of smartphone applications, but one we have yet to use to its fullest potential.

For some of our projects, we have partnered with CitSci.org, which also holds the potential to configure data-collection applications and portals, manage data, and share data with the community. Notably, previously existing templates for project data did not match up with our long-term data-collection protocols, and CitSci staff needed to upload and manipulate the data for upload. We would prefer to build future citizen science projects that are immediately compatible with existing online databases for collection, management, analysis, and sharing of data.

### Data Quality

The Center has established standard protocols for each monitoring project, whether in-house or part of larger local, national, or international citizen science projects. Center staff facilitate training of community scientists, who then can work toward lead roles in research. Center staff ensure both quality assurance and quality control of data collection and entry into databases (chapter 10).

### Data Analysis and Interpretation

Data are analyzed and interpreted using a variety of methods and software. For most projects, data are exported from database storage and summarized using either Microsoft Excel or the statistical software R (R; www.r-project.org). Microsoft Excel has the advantage of familiarity and ease of use for community scientists, interns, and staff. However, data analyses are difficult to trace back over time, and methodology is not inherently stored within the workbook. We recommend that spreadsheet workbooks like Microsoft Excel be paired with data reports (i.e., digital lab notebooks) that document data extraction (Which data were used for the analysis? Were data were transformed for the analysis?) and extensive metadata (column heading definitions, units of measure, statistical methods employed). For long-term research projects at an organizational level, we recommend ensuring reproducibility of research by using statistical software that documents methodology throughout the data-analysis process (from dataset to finished report). We use R because it is a widely used, free, code-based software for data analysis (including tidying and transforming data before analysis) and for data visualization through graphics generation. Through the addition of open-source and field-specific package extensions, analyses and visualizations can be customized to specific reporting needs and can even be exported directly, with text, tables, and figures integrated into finished documents (R Markdown). Although there is a learning curve associated with code-based statistical software, the advantage of documenting data processing and analyses throughout the reporting process offers long-term consistency in the stewardship of data that can be maintained throughout organizational turnover. Numerous free online classes, help forums, and webinars are available, and the investment is also a valuable professional development opportunity for staff, interns, and community scientists.

The geospatial aspect of data is especially important for ecological surveys documenting interactions between organisms and habitats. Because both landscape restoration and wildlife data are constantly created and updated, an archival system needs to be in place. One of our main tools for managing, visualizing, and interpreting geospatial data is ESRI's ArcGIS software, which our community researchers, interns, and staff have been using for over a decade. ESRI provides a full ArcGIS license, including ArcGIS Online and ArcGIS Pro, at a discount for nonprofits. In addition, open-source GIS software such as QGIS (www.qgis.org) and GIS extension packages for R are available. Prior to hiring a GIS specialist, we relied on college interns to create maps and manage data for a variety of projects, including park paths, river bathymetry, and a tree census. Data existed on a shared organization server, locally on a GIS computer, and on the computers of students directly working on projects. Unfortunately, this vulnerable system resulted in the GIS computer acquiring a virus and the Center losing valuable historical data. Additionally, this system was inconsistent; when internships ended or there was an update to software, often no one was available to manage the data or continue license agreements. Without a dedicated data-management system with appropriate metadata, it was sometimes unclear when, how, and by whom data were created.

A particularly helpful student capstone project resulted in a strategic vision for GIS use at the Center, and another internship recommended its implementation using Nolan's six-stage model (Runyard 2005). This strategic vision and a subsequent case study of the Center (Barlow 2006) identified the need for a GIS specialist to acquire data and manage mapping and analysis projects. As a result, the Center sought funding for a full-time GIS and field data coordinator. The position goes beyond traditional GIS activities and requires database management of wildlife and vegetation data, including ecological analysis, interpretation, and reporting. Additionally, the GIS specialist allows the Center to provide meaningful professional development experiences for community scientists, interns, and staff with clear project management oversight and expectations.

### Data Reporting

Project data are reported in the Center's Research and Citizen Science Annual Reviews, in which we provide an overview of the field season, results, and trends. The data are exported from databases and summarized. Data visualizations are created directly in Microsoft Excel, R, or Adobe Creative Suite. Additionally, community scientists and partners are invited to contribute to writing, analyzing data, and visualizing results. The final reports are created in Adobe InDesign and uploaded to the Center's website (https://urbanecologycenter.org) to share with community and professional scientists, community partners, and other stakeholders. Summarizing each individual citizen science project is a time-consuming process, but our goal is to provide space to share the community's annual data-collection and research contributions.

Although platforms like eBird, iNaturalist, and CitSci.org allow for real-time viewing of data through some user-defined analyses, Center staff fill a much-needed role of facilitating opportunities for community members to manage the data they help collect. We provide periodic data workshops on the various data-management systems we use. These workshops helped us understand that not all individuals engaged in our projects want to spend time outdoors and that some prefer crunching numbers indoors. As a result, we developed opportunities for community scientists to visualize our data in graphs and charts, share results with birders who have been collecting data for over fifteen years, publish results in peer-reviewed journals or the Center's blog, and present research at conferences.

We strive to make sure that our data inform not only the management of our restored urban green spaces, but the greater conservation and restoration fields as well. Beyond sharing data with numerous local, national, and international databases, we routinely share data with professional partners, graduate students, interns, and other student researchers. To ensure that our citizen science data are not used for commercial purposes or published without our written consent, we require a signed user agreement before sharing our datasets. We continue to explore better ways to directly share data through our website and other applications to allow real-time community participation in data exploration.

## COORDINATING RESEARCH IN A UNIQUE SETTING

The other major piece of our citizen science "breadth and depth" puzzle fell into place when the aforementioned grant allowed the Center to hire a full-time research coordinator. Fortunately, our previous community engagement enabled us to hire a long-time community scientist who had recently transitioned from another career and knew the ins and outs of the Center's research better than anybody. And because she was active in the organization as a volunteer, she already had an established rapport with fellow community scientists.

### Effective Communication

One of the first lessons we learned was that consistent and reliable community scientists were hard to come by, even though the Center had a volunteer database of over 1,500 people. With the exception of bird projects, turnover was high and new recruitment was inconsistent. Part of this turnover could be attributed to people experiencing the contrast between true field research settings (often a lot of waiting, low action, and empty traps) and media portrayals of scientists radio-tagging cougars and wrestling white-tailed deer to the ground. Regardless, cultivating and retaining volunteers became a priority. Additionally, much of the research coordinator's time was spent on communicating the same information repeatedly to different individuals. Brainstorming with staff and community scientists produced a variety of strategies to improve communication, including the following:

- We created a weekly email newsletter listing times, locations, and contacts for each project. Every other week, we included articles featuring native animals, plants, trees, and so on. The idea was to communicate more effectively and provide space for continued learning. This newsletter replaced individual emails to various people and groups and is now a weekly e-newsletter written in combination with the Center's land stewardship department.

- We added a check-box to the Center's volunteer application for individual research projects to generate project-specific lists of potential community scientists. This check-box eliminated a great deal of "cold call" emails and emails sent to volunteers whom we knew were interested in working outdoors, for example, but not necessarily in helping with a dragonfly survey.

- We began sending new volunteers a personal welcome statement and ideas on how to become engaged.

- We made each project's primary investigator (whether staff, partner, or community scientist) responsible for engaging community scientists through project updates, anecdotes, and photos.

- We added new pages to our website (https://urbanecologycenter.org) to offer a brief history of the Center, an overview of its work, timelines for each project, and so on. Putting more information on the website reduced the need for multiple emails.

- We began conducting regular workshops to recruit new community scientists; update, train, and engage retained community scientists; and prepare for the upcoming field seasons.

- We created the community advisory committee, and an important learning and listening session for department staff led to concrete goals for the future.

### A Broader Target Audience

Early on, we recognized that the majority of community scientists were either middle-class retirees or college students. While both of these groups are valuable contributors to the program, we recognized the need to broaden our reach. One simple fix was to add more weekend and evening programs to target families and community members who work during traditional business hours. Expanding to evenings and weekends works well if there is a structure in place for staff or community leaders to work these hours. An effective way to broaden our reach was to target youths in homeschool or after-school settings and engage them in research (e.g., the Driven to Discover program; see chapter 16). Notably, many of the young scientists who participated in these programs have received awards for their research and spoken at conferences around the country.

### Important Personal Traits

The research coordinator's most important role is as a liaison and supporter of community research. Like a teacher, the research coordinator is required to be "on" all the time or risk losing the audience. If the research coordinator shows disinterest, the volunteers will quickly follow. Likewise, positive energy has the tendency to spread through a group, even if it's raining, the traps are empty, or the mosquitoes and blistering sun seem to be tag-teaming against you. Honesty and appreciation also go a long way ("Wow that was a tough survey—imagine if it only had been one of us"). Any aspects that create and strengthen a sense of community and family are essential, particularly for long-term projects, and any reminders of the contributions of individuals to this team effort go a long way. Sometimes, greeting a person by name and asking a question about their pet or hobby is more important than a certificate to hang on their wall (and this goes for both professional and community scientists). There is a fine line between burnout and deep involvement and appreciation, but a sense of value may be the thing that keeps community members coming back years down the road. Although much of this is anecdotal, the Center is putting a lot of effort toward program evaluation to get a better sense of the elements community scientists value in a program.

## A BALANCE OF POWER AND A RETURN ON INVESTMENT

An important concern that arose from separating the community advisory committee from the professional one was that we were asking community scientists to replace the professional scientists. From our perspective, nothing could be further from the truth. Currently our goal is to create collaborative spaces where professional and community scientists have equal opportunities to contribute to science. We are not asking for one group to replace the other; rather, we are fostering an understanding that academic scientific knowledge and experience are important contributions, but no scientist can work in a bubble. Equally important contributions include anyone's experiential knowledge, connections to community members, careful attention to detail, focus, humor, advocacy, and motivation to understand the world. All of these contributions manifest themselves as big returns for the Center.

It is also essential to remember that as ecologists we are embedded in the social systems of the world, and as such the Center is putting increasing time and effort toward breaking down barriers that promote injustice. In fact, the Center and its research department have continual conversations about ways in which we can promote anti-oppressive policies and procedures (see box 15.1) and hold ourselves accountable at all levels of the organization.

In 2017, after conversations with our commu-nity, the Center officially changed the name of its Department of Research and Citizen Science to the Department of Research and Community Science. The term *community* is more welcom-ing and does not have the potential to insinuate an exclusionary process or the effect of alienat-ing a segment of our society who see the term *citizen* in the context of immigration and natu-ral identity. We followed the lead of Public Lab and were soon joined by other institutions such as National Audubon and the Natural History Museum of Los Angeles County. We have also intentionally replaced the term *volunteer* with *community scientist,* because the overall goal of our program is collaboration. We feel that the term *volunteer* automatically sets a power structure in our favor.

While it is typical for a citizen science program to put a dollar value on the number of hours contributed by citizen scientists for grants or matching contributions, that is the low-hanging fruit. For many years, the Center's communities have invested huge resources toward healing and restoring urban green spaces. The best indicators of how we are doing in this endeavor come from monitoring changes in the local biota—which in itself is a huge undertaking. The most important tool we have for measuring our goals is the iterative proc-ess of adaptive management, where the most effective indicator is the state of the inverte-brate communities most directly affected by the Center's land stewardship work. However, monitoring invertebrates can be an extremely resource-intensive and time-consuming proc-ess that often requires professional input.

Several years ago, funding from the Wisconsin Citizen-based Monitoring Partnership Program allowed us to facilitate a task force of community scientists, whose mission was to help create a feasible and meaningful invertebrate monitoring plan within our means. From this task force, we formed Citizens Researching Invertebrate Kritters Together (CRIKT), a community-driven group that helps the Center better understand the invertebrate commu-nity by building an inventory of invertebrates on our field sites and, more importantly, by helping identify species we should be monitoring as indicators of restoration progress. What seemed like an impossible task is becoming much more effective and allowing us to better serve our mission.

## ADDRESSING SKEPTICISM

In the beginning there was skepticism from many academics, accustomed to working at remote locations, about whether we could make our urban sites with tens of thousands of annual visitors into effective research stations. The concern was that those stations would be totally disrupted by—of all things—people! On the contrary, the dynamic interactions between humans and nature are what make our citizen science research so fascinating. We have a natural ecosystem of which humans are a very big part, and we want to understand how it all works. Additional skepticism from the academic world—and from community members themselves—concerned the ability of community scientists to collect meaningful data and contribute to scientific research. Thankfully, the paradigm is shifting globally in this regard because citizen scientists have repeatedly been shown to produce valid data and carry out high-quality research.

## ACHIEVING FINANCIAL SUSTAINABILITY

At first, the Center's citizen science program was solely grant funded (sometimes referred to as "soft-money funded"). Slowly, the citizen science program began to be incorporated into the Center's operating budget, which included expanding the fund-raising responsibilities from solely the research team to the Center's development and leadership teams. Currently, the citizen science program brings in both earned (20 percent) and contributed (80 percent) revenue. We have experimented with a variety of earned revenue streams, such as workshops for the public, a monthly lecture series, weekend travel programs, instruction for universities, contract work, and national and international travel. Each of these revenue streams resulted in a graph that looked like shark's teeth: each area brought in high revenue one year and low another year, with no consistent growth. This revenue instability did not support sustaining or growing the program, even as the Center expanded and demanded similar work at each branch.

This uncertainty led to a strategy shift. Now we focus on developing just one or two revenue streams at a time and learning how to build and grow the customer base, rather than trying to grow many revenue streams and customer bases at once. For example, the citizen science team was recently hired to establish protocols and conduct research that will help partnering governmental and nonprofit stakeholders track the establishment of a twenty-four-acre urban park from brownfield to prairie, savanna, and woodland ecosystems. This contract represented a significant investment over three years and was recently extended for an additional three years by the stakeholders. Another partnership with local universities and private and nonprofit agencies has the potential for a large, multiyear contract involving community members mapping nontraditional chemicals in their waterways through space and time. Currently, we are working on a half-dozen of these longer-term contracts with partners that engage our community in meaningful research while moving toward a more predictable revenue stream.

## LOOKING AHEAD

Today, the Center's citizen science program works with hundreds of community scientists who contribute thousands of hours of their time with a goal of better understanding the two hundred acres of urban green spaces the Center manages. Our team of scientists (professional and community) contribute to basic science research by developing a deeper understanding of the urban ecosystem, learning about the role of naturalized parks in combating environmental degradation, and uncovering the key elements to sustaining and maintaining green spaces that support a healthy balance of wildlife and recreational space for people. There is increasing desire within urban communities to transform their blighted parks into assets, as we have done in Milwaukee. We believe that our citizen science programs are at the forefront of helping other urban communities understand how urban parks can make the whole community—ecosystem and human population—healthier. In addition to the research more typical at field stations (e.g., bird, snake, soil, plant, and mammal studies), we recently completed a study with the Medical College of Wisconsin finding that kids enjoy playing outside in nature and that it helps them unwind, release stress and anger, and feel healthy (Beyer et al. 2015). But they also have fears about playing outside (e.g., fears of wildlife or strangers). And when they participate in our school programs—which take place in a natural area that is near their homes and their school—they are three times more likely to come back outside of school hours.

Finally, we have learned that you can collect data until it comes out of your ears, but that alone is not helpful. One of our goals is to refine our research so that it can directly inform community members making decisions about the way urban land is managed. We are working with various stakeholders, including Milwaukee County and the Wisconsin Department of Natural Resources, to manage land with an ecological goal in mind *and* to measure progress toward those goals through adaptive management. We feed information to decision makers through our annual reporting and other publications, which highlight progress toward key ecological measures such as water quality, soil restoration, and invertebrate indicators. Additionally, our goal is that all data are open and available to the entire scientific community for both education and management applications. In the long term, we hope to bring our mission to other cities. As one of our core competencies, citizen science is a crucial part of the package.

# SIXTEEN

# Driven to Discover

## A Case Study of Citizen Science as a Springboard to Science Learning

ANDREA LOREK STRAUSS, KAREN OBERHAUSER,
NATHAN J. MEYER, and PAMELA LARSON NIPPOLT

A key benefit of using ecologically focused citizen science as an educational tool for K–12 students is that it typically involves data collection outside, which influences a suite of human behaviors and attitudes (Louv 2005). These behaviors and attitudes include increased knowledge about local ecosystems, increased comfort outdoors, improved attitudes toward the environment, establishment of a lifelong value system of volunteerism and stewardship, a sense of accomplishment and competence, establishment of a lifestyle pattern of physical activity, and positive citizenship and leadership outcomes. Additionally, children who experience nature with a mentor are more likely to take actions that benefit the environment as adults (Chawla 1990).

From a science learning perspective, citizen science engages students and makes science relevant. In fact, direct involvement in projects with real-world applications is more important to building STEM (science, technology, engineering, and math) capacity than classroom or textbook instruction (Swail and Kampits 2004). This involvement applies to both understanding of concepts (Metz 2006, Kirch 2010) and appreciation of scientific practices (Akerson and Donnelly 2010, Berland and McNeill 2010), and is especially important for students with low expectations of success (Hulleman and Harackiewicz 2009). Many citizen science projects enhance participant engagement with scientific topics (e.g., Ferry 1995, National Research Council 2000, Trumbull et al. 2000), promote scientific thinking skills (Trumbull et al. 2000), and lead to increased science knowledge (Brossard et al. 2005, Bonney et al. 2009). However, much of our understanding about the role citizen science plays in education has come from work on adults and not children (but see Juhl et al. 1997, Bouillon and Gomez 2001, Global Learning and Observations to Benefit the Environment 2014).

While citizen science has many educational benefits, there are tensions between using it as a scientific and as an educational tool. At its best, citizen science can help young people build identities as engaged and capable learners (Archer et al. 2010, Tan and Barton 2010), as scientists, and as citizens with the personal capabilities to take part in scientific discourse and decision making (Mueller et al. 2012). As scientific tools, a majority of citizen science projects take a contributory approach (see chapter 1), involving participants as field assistants in scientist-designed protocols. Error-proofing mechanisms ensure data quality, and with sufficient direction and training, volunteers, including youths (Fogleman and Curran 2008), can collect accurate data. This contributory model often does not engage participants in the full process of science. As a result, the contributory model may not be an ideal science education method. However, there is still value in this model, as volunteers are often deeply committed to the outcomes of the research, enjoy their participation, learn about the phenomena under study, and gain confidence. Ultimately, the norms and structures that promote scientific identity through full engagement in the process of science often differ from the norms and structures of many citizen science projects (see also Trautman et al. 2012).

A way to maximize the educational value of citizen science is to promote ownership of scientific investigations by the participants. One way to do this is through co-created projects, which engage participants in most or all science practices (Bonney et al. 2009, Shirk et al. 2012). The Driven to Discover program model is designed to empower adult leaders to mentor youth participants in citizen science. It builds on the idea that participants' citizen science observations often spark their curiosity and lead them to ask questions that could expand scientific knowledge and strengthen their involvement in research. The structured observations of the natural world that are currently embedded in many citizen science projects can provide a springboard for full engagement in the process of science.

## CITIZEN SCIENCE AS A SPRINGBOARD TO INQUIRY

Driven to Discover is built on three steps that lead to three important goals (figure 16.1). First, participants engage in the practices of science, from asking questions to sharing findings. Second, we make citizen science an effective and engaging environment for learning science content. Third, we use citizen science as a springboard to engage in independent research, and thereby help reduce tensions between citizen science as a scientific and an educational tool. To accomplish these three goals, educators and scientists from the University of Minnesota collaborated to conduct a six-year pilot study engaging middle-school-aged youths in citizen science, supporting them as they designed, carried out, and reported on their own research questions under the mentorship of trained adult leaders and scientists. In the pilot study, research teams of youths and adults began by building science skills such as species identification and correct use of science tools like thermometers and binoculars. Then they contributed data to the Monarch Larva Monitoring Program and eBird, asking questions based on observations along the way. Finally, they conducted independent inves-

FIGURE 16.1. Driven to Discover's three-step process. First, participants in Driven to Discover clubs build science skills related to the natural history of target species, taking notes, and using reference materials. Second, they contribute to citizen science by learning the research protocols of the selected citizen science project and recording data accurately. Finally, based on these experiences, participants are prepared to conduct investigations (see figure 16.3). The experience is embedded in a learning environment characterized by support and engagement.

tigations prompted by their questions. All Driven to Discover activities take place within a positive youth-development environment.

### Project Structure

During the development phase of the Driven to Discover project, our goal was to identify factors (e.g., experiences, relationships, and settings) related to participating in citizen science that provoked authentic youth engagement in the science process. We formed research teams composed of youths, adult leaders, and scientists who collected data and contributed to either of two nationally recognized citizen science programs with existing educational materials: the Cornell Lab of Ornithology's eBird and the University of Minnesota's Monarch Larva Monitoring Project (MLMP). This provided the basis for independent research investigations in which the youths formed their own questions and then designed and implemented their own research projects. We studied the factors that contributed to or hindered the research process for the teams.

Our curriculum and training methods were designed to prepare and support volunteer adults as they implemented the program. The curriculum contains background information on the chosen project (eBird or MLMP), instruction on the protocols, lesson plans to help youths build relevant science skills and carry out their investigations, and a Leader's Toolbox that highlights research-based practices for engaging youths in learning teams.

In the second year of pilot testing, we reconfigured the teams, moving the scientists to a supportive rather than leading role. Instead of attending many or most club meetings, scientists served in an on-call capacity, occasionally visiting with teams via Skype or answering questions by email. This on-call arrangement acknowledged the reality of scientists' availability constraints and encouraged team leaders to seek out and make use of other types of community experts, such as Master Naturalists, Department of Natural Resource officials, and other content specialists.

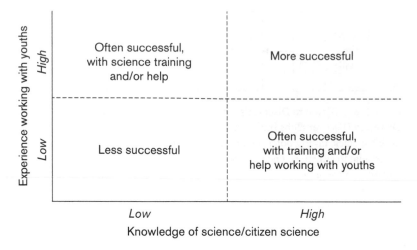

FIGURE 16.2. Adult leader background matrix. Leaders with experience working with youths and a strong background in science were very successful leading citizen science teams (upper right cell), and leaders with either experience with youths or a strong science/citizen science background (upper left and lower right cells) were also likely to be successful with adequate training in the area in which they lacked experience. Leaders with neither experience with youths nor knowledge of science/citizen science had difficulty successfully leading teams.

The program model balances authenticity and engagement to achieve learning outcomes. *Authenticity* refers to providing youths with real scientific endeavors, using real science tools, and connects the experience with prior learning. *Engagement* refers to the ways the club environment provoked the students to learn and practice science, and how it connected them to each other and to caring adults. Balancing authenticity and engagement was critical to ensuring sustained participation. If the experience became *too* authentic, it was boring or too rigid and youths lost interest. If the experience focused too much on engagement, youths became unfocused, the quality of the data they were collecting diminished, and the teams lost the motivation to pursue independent research projects.

## Adult Leaders: Recruitment and Training

We recruited volunteer adult leaders with a variety of backgrounds: middle school science teachers working with youths during summer vacations, Master Naturalists, scout leaders, youth center workers, nature center naturalists, homeschooling parents, experienced citizen scientists, and 4-H leaders. Levels of experience varied widely among these leaders, and we identified two core competencies that adult leaders needed to possess to successfully mentor their youth participants in citizen science and science investigations: experience working with youths in group settings, and knowledge of science or citizen science (figure 16.2). Leaders possessing both competencies (either preexisting or developed in the

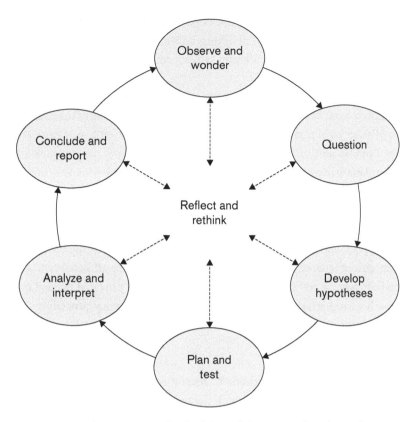

FIGURE 16.3. Science process wheel. This model captures the relationship between the observations made while participating in citizen science data collection and the independent science investigations conducted by youths. Note the positioning of the "Reflect and rethink" stage, which suggests that science is not always linear and that scientists continually reflect on their work and may need to rethink their plans.

program) experienced success with their groups, while leaders with neither competency struggled.

To address the diversity of backgrounds and prepare our adult leaders to work with youth clubs, we provided an annual, two-and-a-half-day training session that covered three topic areas: (1) an overview of natural history and eBird or MLMP protocols; (2) the process for conducting investigations; and (3) practices for working effectively with youths in research teams. Leaders could decide whether they needed to attend each segment, with some opting out of either day 1 or day 3, depending on preexisting expertise. All leaders were asked to attend day 2, where we introduced and practiced the science research process (figure 16.3).

The training focused largely on modeling activities adult leaders would use with their teams. For example, leaders preparing to facilitate MLMP-focused clubs practiced estimating milkweed density and creating a site profile on the MLMP website. Those preparing to lead eBird-focused clubs learned to use binoculars and conduct stationary and traveling bird counts. All leaders engaged in the practices of documenting observations, composing testable questions, forming hypotheses, planning and implementing independent investigations, writing conclusions, making relevant tables and graphs, reporting findings to peers, and reflecting on the experience.

In addition to these core activities, the training included exercises designed to help leaders reflect on their roles as co-learners and mentors. Over the course of the project, we found that leaders of successful research teams not only managed the citizen science and research experiences, but also skillfully addressed the social dynamics of their groups and fostered meaningful relationships with the youths. Our training helped leaders understand the importance of forming effective learning teams, handling conflict, and enabling reflection. For example, we reviewed and modeled the features of a high-quality learning environment (Meyer et al. 2014) and then built in reflection and planning so that adult leaders could consider how to apply these practices with their groups. After the training, leaders prepared to lead summer citizen science research teams. Most leaders were affiliated with an organization that had already gained the community's trust and that provided needed oversight for processes such as registration, health history forms, emergency procedures, and leader background screening.

### Youth Recruitment

Adult leaders used a variety of methods to recruit research teams. In some cases, they worked through an organization with which they were already affiliated, such as a summer camp or a summer school class. Youths (or their parents) chose to enroll in the program specifically for the citizen science program. Other leaders used their existing roles as scout, 4-H, or youth center leaders to introduce the program to a group of youths with which they were already familiar. In these group contexts, the program was integrated into the existing trajectory of group activities. Still other clubs assembled more organically, with parents offering the opportunity to their children and their children's friends.

### Summer Research

Driven to Discover teams met in a variety of locations, including nature centers, school grounds, and local parks. Choosing the meeting location involved considering accessibility, suitability for citizen science data collection, and availability of shade, water, and restrooms. Bird-focused teams were encouraged to meet in a location with at least two different, easily accessible habitat types. Monarch groups needed adequate milkweed, the sole food source for monarch larvae. Both types of teams also needed to consider how to contribute data to a national database. Some teams had Internet access at their meeting site; others met at an Internet hotspot once or twice during the summer to enter their observation data in bulk; some assigned the task to one or two youths to complete data entry at home; and one club leader entered the data herself at home.

We recommended that teams meet for two to three hours per week for ten weeks over the summer. However, some teams met every day, for either half or whole days, for a week or two in a compressed summer-camp format. Many teams continued to meet less formally into the fall as they prepared their research reports. Bird teams were encouraged to meet during times of high bird activity—one club regularly gathered at 6:00 A.M. A few teams met in the evening, depending on the leader's or youth particpants' availability. Regardless of their schedule, teams typically started with an orientation to the natural history of their focal species and corresponding science skills, such as use of binoculars or random sampling methods. Then they collected data for the citizen science project, continuing this each time the club met.

Adult leaders worked to ensure that the data collected by youths were complete and correct, and thus useful to the citizen science program. Bird teams had to ensure that the birds were correctly identified, and monarch teams had to be sure they found the small eggs and larvae on the plants they observed. Leaders handled the need for accuracy in a variety of engaging ways. For example, some leaders created MLMP data-collection zones for pairs of youth observers, while others paired experienced club members with newer members. Some leaders of bird groups assigned one bird species to each youth, who became an expert on that species, while others focused only on counting the birds they observed and not on identifying them.

Throughout the program, teams used an "I wonder" board to note questions that came up, such as "Why aren't we seeing any birds here today?" and "Why does this caterpillar have stripes?" and "Would we see more monarchs if it wasn't so hot out?" and "What are all these little red bugs?" The questions were jotted down on sticky notes and collected. This technique, modified from the Cornell Lab of Ornithology's BirdSleuth curriculum (Schaus et al. 2007), released the leaders from the expectation of having all the answers, while validating youths' observations and questions. The question boards also served as the critical link between the citizen science experience and independent investigations, becoming source material for developing research questions. As the teams became more confident in their citizen science data collection and developed a robust base of "I wonder" questions, they transitioned into independent investigations. With input from their peers, adult leaders, and scientists, youths designed and implemented research projects, either individually or in groups. As these projects progressed, team meetings became less structured, allowing youths the flexibility to work at their own pace.

For teams in which multiple research projects were being conducted, we encouraged Scientist Roundtables, in which individuals or small groups presented their projects and methods to each other. This sharing process helped youths practice phrasing their research questions appropriately, develop workable investigation plans, and collect accurate data. The process also gave youths the experience of presenting their ideas and thinking through multiple investigations as part of a research community.

### Scientist Roles

Early in the project, we paired a scientist—either a University of Minnesota faculty member or graduate student—with each team. Based on feedback from the scientists and adult leaders, we identified points in the process at which scientists were especially beneficial to the teams: learning the natural history of the organism under study, composing testable research questions, and preparing charts and graphs with data. Later in the project, we bolstered adult leaders' skills in these critical areas, and, while scientists remained available to club leaders and youths via email and conferencing software, they were not as closely connected to the teams.

### Research Summit

The annual culmination of the Driven to Discover experience was a fall research summit, which provided an opportunity for youths to showcase their research to peers, family members, and scientists. In addition, the research summit helped youths build an identity of being a scientist by talking about their scientific projects with other youths and adults, and spending time with other science-minded youths and adults. The centerpiece of the summit was a science-fair-style display of research projects where scientists interviewed youths about their research.

## LEARNING ALONG THE WAY

To identify factors related to participating in citizen science that provoked authentic engagement in the science process, a small group of project staff systematically observed several youth clubs and solicited input about the Driven to Discover experience from youth, adult, and scientist viewpoints. The team analyzed data from participant surveys, observations of clubs in action, focus groups, and interviews with adults and youths to synthesize a model of factors that contribute to youth citizen scientists' success in conducting science research and to measure outcomes and impacts on participants (table 16.1 and figure 16.2). Through this process, we identified themes that describe important design elements across three categories:

the setting/situation for each team (the suitability of the program setting for both conducting the citizen science project and engaging in meaningful learning);

the program design/structure (ways that team activities catalyze discovery, activation of prior knowledge, and hands-on skill building); and

team characteristics (characteristics of team members that support learning and teaching about science).

Results of this process were shared with the project team to inform our understanding of the project mechanics and to improve our training and resource materials.

In addition to reflecting on the elements of the model, we worked with adult leaders to refine their practice with youth research teams. In the early years of the project, we saw team dynamics play out in a wide variety of ways and observed that club leaders who possessed the skills to work with youths in group settings also tended to have greater success shepherding their participants through the inquiry process. We strengthened the youth development

**TABLE 16.1**  Factors Associated with Successful Teams

| Factor | Description |
|---|---|
| Program setting | Study site should be ecologically suitable for the intended citizen science project, with ample observable study organisms, and suitable for safely learning about and practicing science. |
| Adult leaders | Attention to creating a positive and supportive learning team removes barriers to full engagement with the science. While experience with the citizen science project and the organism of study is beneficial, leaders should be enthusiastic, curious, and think of themselves as co-learners with the youths, not required to possess all the answers (see figure 16.2). |
| Youth participants | Youths who have chosen to participate are more motivated to engage with careful data collection and the more complex thinking required in the investigation. Summer schedules for youths and their families are widely varied, so club leaders should expect and prepare for patchy attendance, especially in a club whose schedule spans multiple weeks. |
| Program structure | At least twenty to twenty-five hours of contact time allows youths to develop expertise with the citizen science protocols, develop questions, conduct research, draw conclusions, and prepare reports. A showcase opportunity serves as an incentive to complete the research project and provides closure for the research experience. Connecting with local scientists and other experts enriches the experience for youths and adults alike. |
| Communication | Communication with parents and youths should ideally begin from the moment of sign-up with introductions, supply lists, and schedules. It can also include announcements about local events, as well as relevant news and websites. |

aspects of the program by integrating new program features and modeling key practices for adult leaders during the annual training. These research-based program features include ensuring physical and psychological safety, supportive relationships, opportunities to belong, positive social norms, opportunities for skill building, and integration of family, school, and community efforts (Gootman and Eccles 2002). The annual training helped adult leaders recognize their role in setting the stage for positive youth development, creating the conditions for learning, and making authentic inquiry possible.

## A VARIETY OF TEAM EXPERIENCES

Successful research teams represented a spectrum of implementation settings. Our definition of success was based on interviews, surveys of youths and adult leaders, assessments of team research projects, and visits by program staff to team meetings. Among the variety of successful teams, the following four represent the spectrum of implementation models and factors contributing to each group's success (table 16.2).

TABLE 16.2  Example Implementation Models

| Factor | Urban Ecology Center Milwaukee, WI | North Chagrin Nature Center Cleveland, OH | Renaissance Academy near Richmond, VA | St. Hubert School Chanhassen, MN |
|---|---|---|---|---|
| Adult leader | Experienced environmental educator who often partnered with a local bird expert | Experienced naturalist who is also a monarch citizen scientist | Homeschool parent who is also a Virginia Master Naturalist and experienced birder | Middle school science teacher on summer break who is an experienced citizen scientist |
| Youth participants | Five to eleven urban neighborhood youths, ages ten to fifteen, most of whom had participated in other Urban Ecology Center programs | Five boys recruited by a teacher from the local middle school | Ten members recruited from homeschool group, 4-H, word-of-mouth | Twenty or more middle school youths who applied to participate |
| Program setting/ situation | Urban nature center | Nature center | Leader's backyard and community parks | Local arboretum |
| | Implemented program three summers with same youths | Implemented program three summers with same youths | New youths each year | New youths each year (some returning helpers) |
| | Conducted large group investigation | Conducted large group investigation | Conducted individual investigations | Investigations conducted individually or in pairs |
| | Met weekly over the summer | Met weekly over the summer, periodically during school year | Met half days for two weeks summer-camp style; youths completed projects on their own | Met weekly over the summer; some projects completed in fall |
| Factors that contributed to success | Adult leader provided science content knowledge, experience working with youths, a sense of humor, and ability to mentor youth research; he addressed challenges by talking to the parents and cultivating a sense of belonging in the team | Team drew on resources from the nature center, including other naturalists and various wildlife encounters; leader provided frequent reflection opportunities, and investigations followed youths' interests | Leader engaged outside experts and fostered an atmosphere of scientific collegiality; returning teen leaders mentored new members | Leader made connections with science concepts learned in school, helped youths connect with local experts, and facilitated explorations of multiple aspects of the study site, cultivating students' curiosity into full investigations |

## Urban Ecology Center

The Urban Ecology Center (Milwaukee, Wisconsin) hosted several Driven to Discover teams that studied both birds and monarchs. One bird-focused team remained largely intact for three years and was led by an experienced environmental educator and, often, a local bird expert. They met weekly for one to two hours throughout the summer and submitted their observations to eBird on most meeting days. The team implemented whole-group investigations.

The adult leader of the Urban Ecology Center group provided a key blend of scientific knowledge, experience working with youth audiences, a strong sense of humor, and an ability to mentor youth research. He addressed challenges, such as spotty attendance and late arrivals, by talking to the parents and continuing to focus on cultivating a sense of belonging in the team. His supervisor was supportive of the work and recognized the value of the program and the importance of the Urban Ecology Center support structure (see the "recipe" in box 16.1).

## North Chagrin Nature Center

The North Chagrin Nature Center in Cleveland, Ohio, hosted a Driven to Discover team led by a naturalist who had worked at the nature center for many years and had volunteered for the MLMP for over a decade. Her team of five boys met weekly to collect and submit citizen science data for the MLMP or eBird. A teacher from the school and some parents were present for some of the meetings. Each year, the team chose a research project after about a month and worked on this group project for the rest of the summer. The leader described the process by which they came up with a question to study as iterative: they chose a question, then changed the question after more data were collected and after subsequent reflection on questions that would be answerable given their resources.

Meeting at the nature center allowed the leader to tap into many resources. For example, a staff naturalist strengthened team abilities to learn and teach science through sharing knowledge about the local environment, and team members had novel, provocative experiences like holding a tarantula during one of their meetings. The team also experienced challenges, such as a boy stepping on a yellow jacket nest, causing all of the team to be stung. But the leader handled the situation calmly and gave them time to process the event to ensure that the program, and team, would maintain physical and psychological safety.

The North Chagrin Nature Center team members were very much engaged in citizen science. They monitored on their own during breaks, shared questions about what they had observed when they came back together, and developed a strong concern for monarchs through observations and rearing activities. At the end of the summer, the team had a campfire that included their family members and presented their findings at a local "Monarch Magic" event that underscored the value of their studies. Because the youths started back to school before they had finished their research report, the teacher continued to work with them during the school year as they completed the report.

BOX 16.1   Recipe for Success

Tim Vargo, Urban Ecology Center

> 1 cup of flour—This is the science side. We have spent the past ten years building up our equipment and expertise, our protocols and experiences, through our Citizen Science Advisory Council, the Wisconsin Citizen-based Monitoring Network, and other partners.
>
> 1 cup of sugar—This is the volunteers who at first were just "plugged into" projects and are now taking more active leadership in front-end (project development) and back-end (data analysis, visualization, and presentation) roles.
>
> 1 stick of butter—This is the Urban Ecology Staff, with departments acting as one and providing texture and support (volunteer management, development and marketing, environmental education and community programs). This butter became Grade A when we were able to hire a data specialist and a coordinator.
>
> 2 eggs—This is the worldwide movement, led by the Cornell Lab of Ornithology, the University of Minnesota, and others, that we were able to tap into through workshops and conferences, initiatives, resources, and networks.
>
> A pinch of baking soda—This is systems thinking. The Urban Ecology Center has made a strong effort to incorporate systems thinking into our organization, with palpable benefits.

This recipe has been simmering for ten years. It must come to a boil before it reaches a point at which the Urban Ecology Center provides a framework and support for a diverse group of people of all ages to engage in the scientific process, connecting the community to research. Driven To Discover has produced the first popping bubble of this mixture on its way to a boil. This diverse group of fourth- through seventh-graders created their research question, null and alternate hypotheses, and a well-thought-out protocol and then carried out the research, analyzed and visualized the data, put together a poster, hopped in a van, and traveled 330 miles to present their research to their peers at a conference. Additionally, they worked closely with two adult community volunteers who ended up traveling with them. And THEN they came back and presented their poster for the Urban Ecology Center's Research Lecture Series. They stuck around for the presentations of other students and volunteers, asking thoughtful and meaningful questions, and will likely present their poster at one or two more statewide conferences this spring. It is hard for me to believe that we won't see one of these students pursue or strongly consider a career in science.

### Renaissance Academy

The Renaissance Academy was led by a homeschooling parent in Virginia. This adult leader was recruited through the Virginia Master Naturalist Program, had a strong science and natural history background, and was already an experienced birder. The team studied birds all three years. They averaged ten members per year and formed a new group each year (to bring the program to more youths). The leader's own children moved into a teen lead role in later years, providing individualized attention and mentorship and further developing their own expertise.

The first year, the group met for three hours twice a week for five weeks. Because of attendance issues caused by timing conflicts with camps and family vacations, they switched to a summer-camp schedule in subsequent years, meeting for three hours every weekday for two weeks. The group submitted eBird data every meeting to stress the value of contributing to citizen science, and monthly after-school meetings continued into the fall to allow youths to refine reports and assemble display boards for the Research Summit and the Virginia Master Naturalist Conference. This extra time enabled the students to focus on learning and collecting and analyzing data during the intensive camp and gave them time to carefully craft their final reports.

The leader engaged outside experts to share the team's work and help mentor the youths: a local ecology professor, two biology PhD students, and a Driven to Discover project scientist via Skype. They conducted a service project to support local wildlife rehabilitators, who reciprocated by bringing in education birds to allow the students a novel, up-close look at the animals they had been studying. Team members were also invited on a local bird-banding trip.

A key component of the success of the Renaissance Academy team was the strong group context built through a roundtable process for adapting testable questions, conducting peer review of research procedures, and practicing presentation skills. This process created a scientifically authentic collegial atmosphere: all team members "bought in" and learned from each other's research.

### St. Hubert School

The St. Hubert School team was led by a middle school teacher from Chanhassen, Minnesota. Her groups have grown from two youths in the first year to twenty-five, and she created an application process when she could not accommodate everyone who wanted to join the research team. During her involvement in Driven to Discover, there were new youths every year, although former participants sometimes helped. Like the leaders described above, she brought a strong background in natural history to the project. However, her long-term involvement with youths and the MLMP made her unique, as did the large number of youths on the team, the fact that she was almost always the only adult present during the research sessions, and the fact that almost all the youths worked as individuals or in teams of two on their independent projects. This structure required a highly skilled leader and very motivated youths.

The St. Hubert School group met weekly throughout summer at a restored meadow adjacent to a local arboretum, which contained no building or other facilities besides a shaded platform and benches on a hill overlooking the monitoring area. The team began meeting in May and stopped meeting regularly when school started in late August. About halfway through the summer, each young person (or pair) chose a research project. The leader met with them individually, both during their team meetings and off site, and helped them procure materials and refine their questions. In many cases, she helped them make connections with local scientists, sometimes bringing them to the University of Minnesota campus for conversations with scientists. The teacher met with the students when school resumed as they prepared their research posters for the Research Summit and the Minnesota Science Fair.

Because team members had perfected MLMP data-collection techniques, the monitoring segment of the meeting went more quickly, freeing up time for the youths to pursue other discovery activities. They collected and identified bees, looked at milkweed growth patterns, identified earthworms, and surveyed other insect species on milkweed plants. These outside observations provoked topics for many research projects and kept the youths interested.

## BRIDGING CITIZEN SCIENCE AND EDUCATION

The Driven to Discover program uses citizen science to fuel authentic science practice, thus bridging the scientific and educational goals of citizen science. We learned that engagement in science starts when youths are engaged with the team. The formation and presence of a research *team* emerged as a key component to the youth experience. Nearly all youth participants (90 percent) reported that they had attended more than half or all of the team meetings, and nearly 75 percent reported experiencing high levels of enjoyment with the program. By designing and carrying out original projects, youths and adults alike gained a greater understanding of and appreciation for science—and saw themselves as scientists. Evaluation findings mid-project provided evidence that both adult leaders and youths experienced gains (according to self-ratings) in knowledge about the science content and the scientific process; adult leaders also reported gains in their ability to engage youths in science inquiry. Youths reported watching and trying to identify organisms with increased frequency since beginning the project. Further, youths rated themselves, on average, as moderately to very skilled in finding and identifying organisms and habitats, using equipment and guides, and completing data sheets after the project. All these self-ratings increased significantly in comparison to those recorded before their involvement in Driven to Discover.

The Driven to Discover project has produced field-tested curricula and training models that are available for download on our website: https://extension.umn.edu/citizenscience.

# SEVENTEEN

# Challenges of Forest Citizen Involvement in Biodiversity Monitoring in Protected Areas of Brazilian Amazonia

PEDRO DE ARAUJO LIMA CONSTANTINO

The Amazon biome, composed primarily of dense tropical forest, covers more than 49 percent of Brazil. The main strategy for conserving Brazil's Amazonian biodiversity is to create and consolidate protected areas, which currently account for more than 2,000,000 km², comprising 314 federal conservation units (24.5 percent of which do not allow natural resource extraction) that cover 26.1 percent of the biome (1,096,229 km²) and 422 indigenous territories (hereafter "indigenous lands") that cover 23 percent of the biome (1,153,443 km²; ISA 2019). Monitoring of biodiversity is essential for evaluating the ability of the national system of protected areas to conserve biodiversity and improve management of these areas and their natural resources (MMA 2012). To carry out this monitoring, the Brazilian government, through the Instituto Chico Mendes de Conservação da Biodiversidade (ICMBio, the federal agency that manages the conservation units), launched the national program for in situ biodiversity monitoring of federal conservation units (Pereira et al. 2013). Under this monitoring program, the Ministry of Environment recognized the need for social participation, particularly in Amazonia.

The policies on biodiversity contributed to the national policy on environmental management of indigenous lands, designed jointly with indigenous people. This national policy is largely based on indigenous peoples' experiences in developing environmental and territorial management plans of their lands, in which they established a set of internal agreements and recommendations to cooperate with the government and other cultural groups (Apiwtxa 2007). Some of these agreements require data from biodiversity monitoring to improve natural resource management and conservation. However, only a few biodiversity monitoring initiatives involving nonscientists in indigenous lands have been developed within the context of environmental and territorial management plans (e.g., Constantino 2016, Gomes

2017). To highlight citizen involvement in research programs in Amazonian protected areas and provide an example of citizen science projects in the developing world, this chapter will focus on biodiversity monitoring initiatives in protected areas of Brazilian Amazonia.

## CITIZENS IN AMAZONIAN PROTECTED AREAS: SETTING THE CONTEXT FOR BIODIVERSITY MONITORING

Citizen science in Amazonia can differ from projects in the developed countries of the world. Specifically, citizen science in Amazonia is more likely to share commonalities with initiatives in other developing countries in the tropics where people still rely on forest natural resources (Evans and Guariguata 2008, Danielsen et al. 2009). In the case of biodiversity monitoring in Amazonia, citizen science is carried out through the involvement of nonscientists with particular identities and cultures, socially organized in collectives that depend on natural resources from protected areas where they live.

Amazonia's citizens include a number of groups that have struggled for territories and civil rights. Those struggles culminated in (among other processes) the consolidation of socioecological policy of protected lands to allow for sustainable use (Hecht 2011). For instance, in the state of Acre, the term *florestania* was coined to replace *cidadania* and embrace the socialization of forest citizens (Schmink 2011). But unlike participants in other citizen science projects, citizens participating in biodiversity monitoring in Amazonia are individuals organized in family households, often in communities inside or in the vicinity of remote and densely forested protected areas where they depend on the use of biodiversity in various ways and to various degrees (Constantino et al. 2012a, Benchimol et al. 2017, Gomes 2017). In fact, there may be many communities within a protected area. The communities vary in size, but they are usually fewer than five hundred people. Finally, individuals in these communities tend to have low incomes and little formal education (IBGE 2013). One notable location where citizen science occurs is in southwestern Amazonia in the state of Acre.

## MONITORING OF HUNTING IN INDIGENOUS LANDS IN ACRE

In 2004, the Nature Conservancy coordinated a three-year binational project to conserve the Serra do Divisor region on the border of Brazil and Peru. On the Brazilian side, two nongovernmental organizations (NGOs) led the project in the state of Acre: the conservation-focused SOS Amazônia and the indigenous-focused Comissão Pró-índio do Acre (CPI-AC). One of the goals of the project was to collect biodiversity data in order to improve management and conservation in the protected areas located in the Serra do Divisor region, which included two conservation units and eight indigenous lands.

Indigenous peoples were intensively involved in the development of the conservation project on their lands and informed the perspectives on producing biodiversity information. To empower indigenous peoples, CPI-AC used a capacity-building approach by working with indigenous leaders and strengthening community organizations and leaders (Constantino

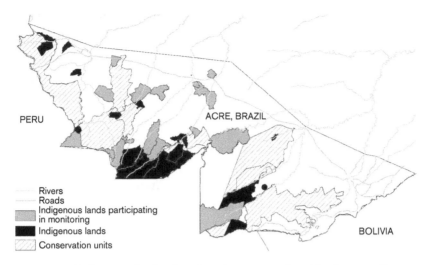

FIGURE 17.1. Indigenous lands where hunting was monitored in the state of Acre, Brazil.

et al. 2012a). Specifically, CPI-AC developed an education program (corresponding to a technician high school diploma) that was recognized by the state, to train indigenous agroforestry agents (IAAs) and teachers to work with their communities in promoting sound management of their own lands (Matos and Monte 2006). All activities related to the conservation project, including the production of biodiversity information through a citizen science approach, became part of the education program.

From the beginning, CPI-AC brought together the stakeholders involved in the management of indigenous lands, including several indigenous leaders. These individuals decided on the guidelines for producing a standardized biodiversity monitoring program that would gather data on the use of natural resources by indigenous communities. Thus, the monitoring of natural resources became a mandatory discipline within the IAA capacity-building curriculum (Constantino et al. 2012a), providing monitoring tools and training to more than one hundred IAAs and their respective communities in fifteen indigenous lands (figure 17.1), far beyond the initial set of individuals who developed the protocols and program.

### The Role of Stakeholders

Indigenous individuals and communities were involved in many phases of the monitoring program, from the design to the use of information to decision making (Constantino et al. 2012a). Leaders, some representatives of IAAs, and indigenous professors helped design the methodology. The IAAs were responsible for coordinating the monitoring in their own villages by organizing data collection, collecting data, training other citizen scientists, conducting simple data analyses of their villages, leading local meetings to share results, and promoting community involvement. The IAAs were paid by the state through a scholarship, and there was a formal commitment to contract them as practitioners. Notably, however, payment was tied to an IAA's environmental extension work with communities, the monitoring being only one activity and not a requirement for being paid. Therefore, although all IAAs were trained and responsible for coordinating the monitoring in their villages, their own and their community's engagement was voluntary, as they could choose not to participate in the program.

A team of two CPI-AC biologists were responsible for leading the design of the monitoring program and protocol, promoting capacity building in communities and among IAAs, centralizing and storing data, producing educational materials and documenting results, and supporting interactions with scientists and other stakeholders. Collaborating scientists helped design the monitoring protocol, helped analyze the data, and provided insights regarding management.

### Selection of Participants

The IAAs were nominated by their communities to participate in the education program. The criteria for nomination included a commitment to participate in capacity building and to carry out environmental extension activities in the village. Given the gendered division of labor in Amazonian indigenous peoples, IAAs were mainly young male volunteers. Communities also had the power to fire the person in their IAA position. As a result, there was a continuous collective evaluation of the IAA's work that was reported to CPI-AC by community leaders.

Since the monitoring program was tied to the activities of the IAAs, they were automatically in charge of local coordination. Depending on the internal arrangements in a given village, the IAA recruited literate community members (students in the village school or hunters) who were able to use tables, read maps, and input hunting data. The involvement of IAAs in long-term research was facilitated by their prior education, their interest in conducting research with their people or on their lands, and their having presented a research monograph as a requirement of their graduation from an institute of higher education (Monte 2000).

After the first community meetings to explain the program, hunters volunteered to provide data without restriction and with the commitment to provide data every time they went out hunting. Other community members and leaders participated in several other activities related to the monitoring program, such as collective data analysis and interpretation and reporting results during community meetings.

### Design of the Monitoring Program

The program was initially designed to gather continuous data on the use of selected biological resources: wildlife, fish, palm trees, and timber. These natural resources were chosen by the indigenous people because of their need for basic subsistence. With the exception of palms, the monitors used standardized forms to record data on extraction from families willing to provide information every time they engaged in hunting, fishing, or logging. This methodology was easily implemented, appropriate to the dynamics of the communities, and not considered time-consuming. Indigenous representatives understood that the methods used to collect data on populations or stocks of game animals, fish, and timber were inappropriate in this regard and would require more effort.

During the first three years, approximately forty-five communities (around 40 percent of the trained communities) in eight indigenous lands (more than 50 percent of those initially involved) conducted high-quality monitoring of hunting (Constantino et al. 2012a). However, monitoring of fishing and logging was restricted to a few families in a few villages. Therefore, hunting became the focus of monitoring efforts by indigenous people, CPI-AC, and associated researchers. One of the reasons identified for the communities' focus on monitoring hunting was that they considered it too time-consuming to monitor all the natural resources initially considered. Because game meat is the main source of protein and has a central place in the organization of these societies—hunting and wildlife thus being a recurrent theme of conversation for all these indigenous peoples—they selected this target for long-term monitoring. Also, many villages had experienced variation in game species availability over the past twenty years for several reasons, resulting in a need to shift their economy (Constantino 2016). The monitoring program ended in 2011.

### Capacity Building

At least seven courses for IAAs were held in the Centro de Formação dos Povos da Floresta school in the capital of Acre; more than twenty workshops for IAAs and community representatives were held in the indigenous lands; more than twenty months of follow-up visits in villages were conducted by CPI-AC technicians; and more than thirty IAAs have traveled to visit other monitoring projects in Brazilian Amazonia and abroad during the six years of the monitoring program (figure 17.2). In parallel, depending on how the IAA decided to run the program in their village, they also trained backup monitors, either informally or in the local schools, with support of the indigenous teachers. As mentioned above, biodiversity monitoring became a discipline in the formal IAA curriculum that was designed to train individuals in data collection, simple analysis, data interpretation, and the use of specific tools needed to carry out these actions (e.g., indicators, spreadsheets, graphs, maps, GPS; figure 17.3). In addition, a conceptual part of the curriculum focused on the reasons for scientific monitoring, the context of the program within larger social and ecological processes, the rationale and importance of biodiversity monitoring, and the use of information in decision making.

### Data Collection

In each community, the IAA was the main person responsible for collecting hunting data, but they often recruited assistants to help. CPI-AC delivered enough monitoring books with data forms and village maps for one year of monitoring. Specific information on the data forms included where animals were hunted, details about the animals (e.g., species, weight, sex, and estimated age), and details about the hunting trip (e.g., the technique and weapon used, duration of the hunt, location where the animal was killed, and distance of the killings from the village). Monitors adopted a variety of mechanisms to record data from all hunting events (figure 17.3), including visiting the hunters' households on a daily or weekly basis, recording data during weekly community meetings, and (most frequently) recording data at collective meals. Conversations about hunting and the animals eaten are common during these collective meals, and monitors wisely used this opportunity to collect data. Indigenous people also participated in the production of supplementary hunting information that supported the analysis of monitoring data. Some of them used a GPS device to gather data on the exact location of hunted animals and landscape features important for hunting (e.g., salt licks, fruiting trees, hunting trails, and campsites) and also to map the limits of the hunting areas.

### Data Ownership, Storage, and Management

Initially, all the completed paper forms were delivered to CPI-AC during the capacity-building events or follow-up visits, digitized, and centrally stored as a simple spreadsheet in the main office. Only CPI-AC staff could manage the data. Original hard-copy forms were stored in the CPI-AC office. As the program evolved, communities requested to keep their original data in the villages and only send digital copies to CPI-AC to be digitized in a central spreadsheet. Storage conditions in the villages are awful, especially with regard to paper, given the high humidity in the Amazon. However, indigenous people find it important to have their original data in case they need to run additional analysis and have quick results. Data use was regulated by CPI-AC and the indigenous people, but the Nature Conservancy had a copy of the dataset, which could be used only if authorized. Collaborating researchers had authorization to analyze the data. If the results were published, the researchers were expected to provide the published findings to the communities.

### Data Analysis

Data analysis occurred with involvement of indigenous people at the scales of the village, the indigenous lands, and the region. Ideally, every month, after running a simple analysis, the IAA would present the summarized hunting data during a community meeting. The IAA indicated the total number and weight of hunted animals by species using tables, showed the variation over time using figures, and identified the distribution of hunted animals on a map (figure 17.3). Each year, the community leaders, teachers, IAAs, and members involved in the monitoring met to analyze the summarized information from the villages of the indigenous lands using the same simple analysis employed at the village level. During these annual meetings, there was no involvement of the CPI-AC or other scientists. The latter, however, performed analysis at the regional level and returned the results to the communities to discuss the emerging patterns and processes, resulting in the joint publication of the research in scientific journals (Constantino et al. 2008, 2018; Constantino 2015, 2016, 2019). Nevertheless, only a small proportion of communities actually analyzed the data locally and regionally, because most of them did not participate in the data analysis component. Although training for data analysis could be simple, the indigenous people are not used to producing or reading graphs and tables. Even the most literate individuals required another whole series of courses to be able to participate in these activities.

## Management and Decision Making

The results of the simple analyses performed locally in the villages were the only findings used to support management actions. Using these local results, management decisions were made by communities with the guidance of village leaders and IAA. Some communities decided to either increase or ban the hunting of certain species depending on observed trends, as well as to maintain or restrict hunting in certain management zones, in

FIGURE 17.2. Indigenous agroforestry agents (IAAs) during capacity-building courses. Credit: photo by Pedro Constantino.

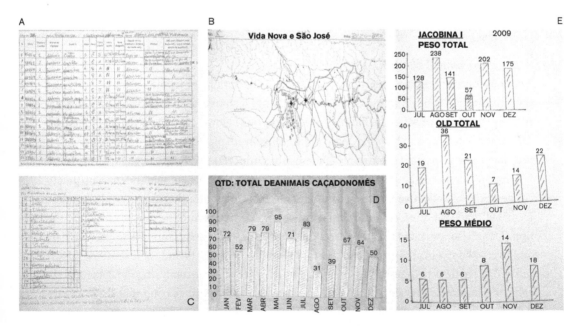

FIGURE 17.3. Indigenous agroforestry agents collected hunting data on (**A**) spreadsheets, (**B**) maps, (**C**) analytical tables, and (**D, E**) graphs. Credit: photo by Pedro Constantino.

agreement with the norms specified in the environmental and territorial management plans. Furthermore, the results led to proposals to intensively manage some species, such as agouti (*Dasyprocta* spp.) and collared peccary *(Pecari tajacu)*. Inside indigenous lands, only indigenous people had the rights to make management decisions, while technicians and scientists provided additional information and facilitated the debate as needed. Political decisions based on the hunting data and the monitoring program were rare. Nationally, the monitoring of hunting in indigenous lands was used to suggest guidelines to improve the criteria used to identify the limits of indigenous lands on the basis of indigenous hunting dynamics and territories (Constantino et al. 2018), as well as to provide guidance on the design of aspects of nonscientists' involvement in the Ministry of Environment's national program for in situ biodiversity monitoring and of the capacity-building program to support it (Santos et al. 2014).

### Evaluation

The monitoring program was evaluated using several approaches. Data recording and analysis were evaluated by CPI-AC during capacity-building events and follow-up visits, when technicians tried to clarify any imprecision with the IAA. Based on these evaluations, capacity-building tasks were created to minimize future errors. During these events, the protocols and methodology were also evaluated in order to make any adjustments. CPI-AC technicians checked the collection of hunting data in the villages by following hunters during hunting trips and by double checking random data that the hunters provided. The IAA's performance as a monitor was evaluated by the community, which reported to CPI-AC staff during follow-up visits. CPI-AC also evaluated the IAA's performance by analyzing the data collected and analyses conducted.

### Challenges

#### Involvement and Motivation

The monitoring program initially had ambitious expectations of involving individuals and their communities in gathering data on many natural resource items and using the information for different purposes, including formal education. However, various stakeholders had different expectations, which may have reduced involvement. First, some IAAs argued that they had more important extension work and other important activities (e.g., farming, hunting, fishing, construction) than collecting hunting data, while others said they saw no practical personal or collective meaning in producing information on hunting. Similarly, some IAAs believed that hunters already knew about wildlife trends and the causes of their variation. IAAs also indicated that they faced difficulties in convincing hunters to provide data and in involving families in the monitoring. Second, the hunters and their families simply refused to participate in providing hunting data or collecting and analyzing the data. Specifically, the hunters distrusted the reasons for monitoring wildlife, fearing that the data produced would result in retaliation, hunting bans or control, internal disputes with the family of the IAA and other leaders, and disagreement over paying only the IAA when many families were participating for free. Third, interest in maintaining the monitoring program decreased in the CPI-AC as the end of the main funding neared and after a complete personnel change in the NGO team that designed and helped implement the monitoring program. In fact, no capacity-building events on hunting monitoring were conducted after 2011, despite the fact that high-quality data from the project had produced a regional assessment of indigenous use of wildlife (Constantino 2016) and were available to evaluate the agreements present in the management plans of the indigenous lands that CPI-AC helped develop. Fourth, the scientists who assisted in the project lacked incentives to continue because the results of the program were not being used by the NGO, the indigenous people, or the state government. Only recently have other scientists contacted CPI-AC to express interest in using the monitoring methodology to explore new participatory monitoring technologies.

### Data Recording, Management, and Evaluation

Several challenges arose in regard to the data. First, recording of data was problematic in that common Portuguese or indigenous names of species were used to fill out the forms instead of the standardized names. However, this naming problem was easily solved through continuous follow-up visits that created a common understanding of the terms. Second, biases were apparent in the data, in that larger and more important animals were more frequently recorded than smaller and less preferred animals. Third, there was variation in the precision of data among monitors, which could be a source of uncertainty. New technologies such as smartphones could reduce uncertainties in recordings, although some adjustments would be needed to avoid excessive cost. Moreover, such technologies would probably speed the feedback of results to communities. Fourth, the accumulation of paper data forms to be digitized and the absence of an appropriate database to store monitoring data made it difficult to quickly produce reports that could be returned to the communities. Likewise, the absence of automatized mechanisms to evaluate data quality resulted in manual evaluation of more than ten thousand entries of hunted animals with all associated data in the spreadsheet. Fifth, although the dataset could be accessed by virtually anyone, a person would need to request the latest version of the dataset from the few indigenous leaders who kept the digital file, from CPI-AC, or from the collaborating scientist. In order to solve this issue, CPI-AC installed infrastructure to store the data in a database together with the GIS information of the indigenous lands.

### Use of Information

Using the monitoring information was a great challenge. Even where indigenous people had autonomy to manage the resources in their lands, and there were local collective agreements on wildlife management that required monitoring information (i.e., AMMAI-AC 2007), information was rarely used to support decisions. At larger scales, monitoring information was used to produce scientific knowledge, but so far the information has not been used to support decisions on natural resource management.

## Lessons Learned

A number of important lessons were learned through the hunting monitoring program. First, the lack of support from CPI-AC and collaborators culminated in an end to the hunting monitoring program in 2011, although some villages kept collecting and analyzing data independently until 2012. No villages monitor the use of natural resources at present. However, the main objective of educating people in research and other disciplines was achieved.

Second, the emphasis on education, the linking of monitoring with the capacity-building program, and the strengthening of collective action were important strategies that motivated and maintained individuals' and communities' participation in the program. Although an important aspect motivating IAA participation was paying for their monitoring services, it clearly did not guarantee long-term, high-quality involvement in different phases of monitoring. A motivation for community involvement was the possibility to run simple data analysis and produce useful information with the data a community collected. The involvement of village schools and leaders, data recording and debate during collective meals, local meetings for analyzing and evaluating data, interchange among communities of different indigenous lands, and the participation in national and international political and scientific events were some of the mechanisms adopted that led to intensive citizen participation in the monitoring.

Third, the monitoring program contributed to the culture of collecting systematic data both among the indigenous people and in CPI-AC, and there were indications this culture would evolve to include other aspects of biodiversity monitoring once information started to support management decisions. Specifically, the structure to implement participatory monitoring is now in place, with local representatives aware of their responsibilities and communities aware of the purpose of keeping track of biodiversity data. The communities of Praia do Carapanã indigenous lands, for example, decided to complement the information from hunting by monitoring game species in their forests through line-transect surveys after they realized that the reduction in their success hunting howler monkeys *(Alouatta seniculus)* might be a consequence of the reduction in the animal's population and not of hunters' bad luck (Constantino et al. 2012b).

Fourth, the experience with monitoring natural resource use led CPI-AC to raise significant funding to develop participatory monitoring of land-use change in the villages, based on remote sensing and ground truthing, in which the IAAs are trained to conduct GIS analysis and collect field data.

As Brazil implements its new national environmental management policy, there is an increase in management plans for indigenous lands and a growing need for long-term biodiversity information. These needs may create a demand to restart the monitoring within the indigenous lands in Acre, as well as to start similar ones in other states of Amazonia in the future. The lessons from the hunting monitoring project, as well as those from other monitoring initiatives in Amazonia, have already contributed to the design and implementation of the national program of ICMBio to manage conservation units.

## CHALLENGES IN OTHER CITIZEN MONITORING PROGRAMS IN AMAZONIA

As of 2012, there were at least eighteen initiatives of biodiversity monitoring with nonscientist participation in Amazonia. In such initiatives, although the general context is the same as for the monitoring program described in this chapter, indigenous lands are governed differently than conservation units, in which use of natural resources is regulated by the state, and differently than areas designated for "sustainable use," which are co-managed by the state and local communities. This difference brings powerful stakeholders to the table, which can result in different challenges for citizen participation in monitoring.

The involvement and constant presence of conservation unit staff and (to a lesser degree) of the local partners, such as research and capacity-building organizations, is essential to motivate and maintain nonscientists' participation. However, conservation unit staff are extremely reduced—frequently consisting of two people to coordinate and manage an entire protected area—and have elevated turnover rates, and partner organizations usually have other commitments. Therefore, people, time, and presence in the communities are real constraints to the participation of local people in monitoring.

Although various mechanisms have been adopted to pay for monitoring services, this does not guarantee the proper execution of monitoring and data quality. Moreover, because of legal constraints, it is currently almost impossible for the state to pay monitors. On the other hand, financial rewards have not been an issue in monitoring designed to inform management of commercial species, where management norms are agreed upon by the local communities and the state. For example, information from pirarucu *(Arapaima gigas)* monitoring was used to set fishing quotas, and fishing profits were distributed among the fishers and the monitors. In such cases, the association of monitoring with a natural resource of major importance to a community, particularly a community that has management rights, significantly increases the interest in and quality of local participation.

Providing sensitive information is another important aspect that has challenged monitoring initiatives. In some cases, hunters were supposed to report the hunting of endangered species to organizations that might eventually help regulate that hunting. In cases like this, stakeholders have to build a strong relationship based on trust, often ensuring that monitors and resource users remain anonymous.

Finally, a great challenge experienced in some initiatives was poor feedback of information to the communities and a lack of support in management and political decision making. One reason for this was the accumulation of a large amount of data in paper forms and the absence of a data management system that could properly store the data and quickly provide results of simple analysis. Therefore, most of the data from all these initiatives have not been analyzed.

# EIGHTEEN

# Documenting the Changing Louisiana Wetlands through Community-Driven Citizen Science

SHANNON DOSEMAGEN and SCOTT EUSTIS

Community science and participatory environmental health monitoring have rapidly developed over the past decade, with groups focusing on tools and technologies ranging from community air and water monitoring tools to wearable technologies tracking the ecosystem changes that affect our personal health. Public Lab (https://publiclab.org) is one such group that arose in response to the need for greater environmental monitoring. Specifically, during the first week of the Deepwater Horizon oil spill, it became apparent to residents of the Gulf Coast that obtaining information that was verifiable and trustworthy would be an issue. British Petroleum (BP), the corporation responsible for both the pollution and the cleanup, heavily controlled the official response. People from the growing Grassroots Mapping community joined with the Louisiana Bucket Brigade, a nonprofit that monitors industrial facilities in the state, to embark on the first large-scale instance of a Public Lab project. From across the Gulf Coast states of Louisiana, Mississippi, Alabama, and Florida, residents launched tethered balloons and kites 1,500–2,000 feet in the air with simple point-and-shoot cameras attached, effectively capturing over a hundred thousand images of the Deepwater Horizon disaster as it progressed. Using MapKnitter (https://mapknitter.org), an open-source, cloud-based software program developed by Public Lab, the community created hundreds of maps of the coastline that can now be found in Google Earth as historical and primary layers. Images captured during this time were used to describe alternative public stories through large media outlets (see https://publiclab.org/media).

Today, Public Lab is a global, open community, supported by a nonprofit, that cooperatively creates and develops open-source hardware and software tools for environmental monitoring and analysis. The nonprofit supports chapters and organizers working worldwide in partnership with community organizations to use tools and methodologies developed by the community. When creating the structure of Public Lab, the founders took a step back and identified several key issues that they were interested in addressing as the organization was formalized. For instance, because many environmental monitoring tools are developed at a price point for corporations, government agencies, and research institutions, but not at a price accessible to everyday people, the founders sought to create an open space where people could come together in a model of community science and develop monitoring tools.

Since its creation, the Public Lab community has developed into a broader and more structured organization. While still focused on developing low-cost, collaborative tools and techniques for environmental monitoring and analysis, the community has improved how information is communicated and shared. For example, our Research Notes weblogs introduce newcomers to the basics of scientific practice and how to share their knowledge. These weblogs are written from prompts that encourage individuals to introduce a topic, describe their methods, post their results, and discuss them. In addition to the online content, Public Lab has also developed low-cost monitoring tools that individuals can purchase and build themselves, such as the aerial mapping tool kit. Public Lab also encourages members of its community to produce their own kits and sell them through the Public Lab website.

While Public Lab is a global community, a large focus remains the Gulf Coast, particularly the Mississippi River Delta. As such, the delta provides an important case study in demonstrating the value of Public Lab in community-based monitoring through the platforms and tools used and the community it represents.

## THE NEED FOR MONITORING: A BRIEF HISTORY
## OF THE MISSISSIPPI DELTA

The Mississippi River Delta is a large and unique estuary, contained within the state of Louisiana, that provides a number of important ecosystem services (Barbier et al. 2008, Batker et al. 2010). For example, the delta is a critical hub for bird migration as part of the Mississippi Flyway, contains the largest contiguous bottomland forest on the North American continent, and produces a quarter of fisheries landings by weight in the lower forty-eight states (Batker et al. 2014). These and other ecosystem services provided by the delta have been calculated to represent $12–47 billion annually (Batker et al. 2010). Furthermore, the delta is the site of historical and modern industrial food and fuel production and transport for the United States. Largely as a result of past and present industrial use, the delta has lost almost 1,900 square miles since 1932 and loses roughly an acre of land to open water every hour in its current state of degradation (Couvillion et al. 2011). This land loss amounts to 90 percent of total wetland loss in the United States.

The deterioration of the delta landscape—recognized since the rate of loss peaked in the 1970s—was a major factor in the flooding of New Orleans in the aftermath of Hurricane Katrina in 2005. The first effort to address this deterioration was the Coastal Wetlands Planning, Protection and Restoration Act (CWPPRA; also known as the Breaux Act), passed by the U.S. Congress in 1990. The CWPPRA program has spent millions of dollars and nearly thirty years on ecosystem-restoration project planning, engineering, construction, and evaluation. Although successful, the program has been criticized for being of insufficient scale and bureaucratic priority in the wakes of Hurricanes Katrina and Rita. In response to the levee's structural failures during Katrina, and a subsequent engineering report (National Academy of Sciences 2009) outlining various system failures that led to the disaster, a methodology was developed to evaluate the land for flood protection and reconstruction of ecological processes (Coastal Protection and Restoration Authority 2012).

Subsequently, the State of Louisiana created a bureau in the Coastal Protection and Restoration Authority (CPRA) to oversee the complicated bureaucratic and physical systems necessary to minimize flood risks on the low-lying coast. The CPRA is charged with monitoring the delta ecosystem, evaluating the hundreds of proposed restoration projects accumulated over the past twenty years by the CWPPRA, coordinating this restoration with levees and other forms of flood risk reduction, and implementing an integrated suite of priority restoration projects that could fall within a $50 billion budget and a fifty-year time frame. Approximately half of this budget ($25 billion) is dedicated to ecosystem restoration, while the other half is allocated to flood risk reduction through structural approaches (e.g., earthen and concrete levees, locks) and nonstructural approaches (e.g., flood proofing and elevating homes, buyouts of coastal communities; Dalbom et al. 2014). In essence, then, these actions collectively fall into the Coastal Master Plan, which puts ecosystem services on government ledgers.

Because Louisiana derives a large amount of revenue from oil and gas, the Coastal Master Plan includes "support for the oil and gas industry" as a planning goal (notably, however, the plan outlines no role for the industry in paying for coastal damages it may have caused). Furthermore, the Coastal Master Plan's call for ecological restoration consists primarily of mechanical manipulations of the delta landscape, rather than traditional land preservation or conservation and management efforts. Following the Deepwater Horizon disaster, this focus on mechanical manipulations came to a head with the proposed installation of a high, thin wall of sand segments, or berms, around the entire delta, ostensibly to shield the interior marshes from heavy oiling. The governor of Louisiana moved forward with the sand berm proposal, which was to cost hundreds of millions of dollars and have no planning input from government, NGO, and university experts who had been studying, planning, and designing barrier island restoration projects on the Louisiana coast for decades. In addition, aerial imagery of bulldozers piling sand across the shallow waters of the Chandeleur Islands as waves washed the berm project were front-page news in regional papers. After objections from NGOs, restoration experts, and federal regulators—and media attention—the plan was altered from berm construction to barrier island restoration (which had already been planned). The state relabeled the response "Berms to Barriers" and currently lists its action as a great success (Coastal Protection and Restoration Authority 2014, Condrey et al. 2014).

As the Berms to Barriers program demonstrates, there is a great need for attention to both policy and environmental monitoring. In Louisiana, there are insufficient resources to monitor pollution sources (Environmental Protection Agency OIG 2011), and thus nonprofit organizations have played an important role in documenting environmental problems, such as leaks and spills after seasonal hurricanes (e.g., Gulf Monitoring Consortium). Today, low-cost environmental monitoring tools are important both as a mechanism for accountability (with regard to the Coastal Master Plan) and as a means to educate and engage affected people.

## CITIZEN MONITORING: THE BARATARIA MAPPING PROJECT

In 2011, members of Public Lab's Gulf Coast chapter traveled to the marshes of northern Barataria Bay, Louisiana, with Dr. Alex Kolker from Louisiana Universities Marine Consortium. Dr. Kolker was conducting research into erosion rates of marshes oiled by the Deepwater Horizon disaster across twenty sites ranging from *no oiling* to *heavy oiling*. During several of these outings, Public Lab collected aerial images and created maps detailing bare, gray-brown marsh platforms. In addition, other Public Lab teams collected photos showing the extent of the brown oil smothering green marshes, as well as the relative dysfunction of the white lines of sorbent boom in open bays like Barataria, which were shared with local residents and national media (Dosemagen et al. 2011). These photos provided baseline assessment of damage from the spill at several locations, allowing Public Lab the opportunity to revisit and monitor the damage and cleanup over the long term.

In 2014, with financial assistance from the Patagonia clothing company, Public Lab's Gulf Coast members sought to repeat the survey in a more systematic fashion. Because different individuals were involved in the 2014 survey, they first had to learn the aerial mapping methodology. Public Lab staff and organizers led local field demonstrations, and the larger Public Lab community led map outings at Public Lab's annual "Barnraising" meeting. Members learned and taught by asking questions and reading answers via the Public Lab and Grassroots Mapping listservs, and by reading instructions and watching instructional videos in Research Notes created and shared via the Public Lab website. A core group of individuals focused on map making (i.e., MapKnitters) met every other week to plan the outings; to learn and teach one another how to use the Public Lab website, GIS, and MapKnitter software; to divvy tasks among members; and to write Research Notes themselves.

For each mapping outing, participants surveyed three to five study sites. However, due to focusing errors, not all individual mapping flights were successful. Both balloons and kites were deployed for mapping (figure 18.1), depending on weather conditions. Generally, if winds were less than five miles per hour, a balloon was deployed, whereas if they were more than five miles per hour a kite was used to lift the camera. Flying the balloon was attempted first. If the balloon was unstable in the air, due to variable or high winds, the kite was flown. On two occasions, because of a lack of helium, the kite was lifted in the absence of wind by propelling the boat across the water. Lift of the balloon or kite was measured in kilograms, using a simple suitcase scale hooked to the string. Depending on the lift of the balloon or kite, different rigs (picavet, soda bottle) could be used, the soda bottle rig being the lightest. The camera of choice for this project was consistently the Canon Powershot

An Illustrated Guide to
## Grassroots Mapping *with*
## Balloons and Kites *v. 2.2*
*To see all the latest contributions and video instructions,*
*visit http://publiclab.org/tool/balloon-mapping*

*Do you want to make maps? Do you need satellite images but can't afford them? Do you want to see your home from above?*

*Follow these instructions and you can, for as little as $100!*

This work is licensed under a Creative Commons Attribution ShareAlike 3.0 License.

a large kite - 1m² or more

One 2 meter-wide weather balloon or 5 chloroprene "cloudbuster" balloon. Chloroprene balloons are more durable.

or 2 84" mylar sleeping bags

1000m 5kg nylon string for balloons

1000 m
5 kg

30kg+ strength nylon string for kites

1 m
80

digital camera with continuous mode + 4 gb or larger memory

heavy work gloves

2 L

plastic soda bottle

~200g

scissors

80 cubic feet or 1.5 cu. meters of helium

rubber bands

duct tape, gaffer tape is best

FIGURE 18.1. Public Lab's illustrated guide to grassroots aerial mapping.

1400is, on continuous-shot mode. Pictures were taken about every second with a knot pressed into the camera trigger by a rubber band wound tight around the trigger, secured with gaffer's tape. This method enabled the capture of thousands of images as the camera flew at hundreds to thousands of feet above the wetland landscape.

After ten to thirty minutes of flight, depending on field conditions and site quality, the balloon or kite was retrieved, which usually took thirty minutes. Once retrieved, the camera was checked to see whether it was still operating. Photography was reviewed on the camera display to determine whether the target area was captured and whether the images were in focus. About one flight in eight returned blurry photos. One flight had a camera lost to the bay as the balloon was retrieved, but a spare camera and rig were available to complete the trip.

From the images taken on each trip, a group of five Public Labbers stitched the maps together in MapKnitter. There are many challenges involved in mapping a landform that is disappearing, but methods were developed to overlay 2014 wetlands photography onto U.S. Geological Survey base maps using features like oddly shaped lakes and the confluences of bayou channels. Although many Public Labbers have started maps, stitching has been completed by four Gulf Coast members with GIS training, one Public Lab staffer, and another Public Labber with cartography training. Once maps were completed, they were exported from MapKnitter into QGIS for final analysis and display. This analysis included a measure of how much land was lost and gained from 2010 to 2014.

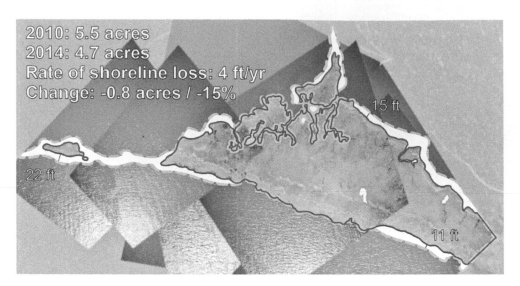

2010: 5.5 acres
2014: 4.7 acres
Rate of shoreline loss: 4 ft/yr
Change: -0.8 acres / -15%

15 ft

22 ft

11 ft

FIGURE 18.2. Map of Barataria Bay Heavy Oil Site 2.

### Results of the Mapping Project

Training participants is a part of implementing the mapping technique. For each outing, about a hundred people were invited using the Public Lab email list, Gulf Restoration Network's contact list, and personal contacts via email and phone calls. Over the course of 2014, a total of six field excursions were taken, consisting of twenty-two different participants. Participants traveled to Barataria Bay in recreational fishing boats, some hired and some captained by participants. Of the total of twenty-two people who participated in mapping trips, most were trained in one or more aspects of the map production process. Flying and launching the kite or balloon was more appealing to the majority of participants (sixteen and fifteen individuals, respectively). Almost half (ten) prepared the camera for photography and attached it to the kite or balloon line. The conceptual, analytical, and logistical tasks, however, attracted fewer participants, with five stitching a map, two annotating a map, and four organizing an outing.

More than ten maps of northern Barataria Bay shorelines were produced by kite and balloon, four of which have been analyzed in QGIS for shoreline change (http://publiclab.org/maps/tag/barataria). The minimum resolution of the maps is 2.9 cm/pixel, the maximum is 11.77 cm/pixel, and the average is 6.98 cm/pixel. All maps extended a minimum of three hundred meters across a designated shoreline. The extent of maps varied from thirty acres, for a very low-altitude kite map at Barataria Site BL1 (Long 2014a), to a 350-acre map from a high-flying balloon at Pelican Point (Long 2014b).

Figure 18.2 shows an example of the maps the participants produced. This site experienced some of the heaviest oiling in 2010–11 and is where some of the more experimental cleanup and response techniques were deployed, including vacuuming of oil and raking.

The peninsular shape of this landmass exposes it to erosional forces from all sides, and from 2010 to 2014 the site lost nearly 15 percent of its landmass. Notably, the island is composed of both vegetation-free areas (black) and areas where *Spartina* plugs have been replanted (dark-colored). The small island and the larger spout feature were connected until 2014, when erosion created a new island and the boundary between Bay Jimmy and Barataria Bay disappeared.

### Importance of the Mapping Project

Public Lab's aerial photography method provides an important tool for motivated people to document shoreline retreat of oiled marshes. One advantage of the suite of open-source map publication tools is that the maps and the data generated are more widely available for review than standard aerial-photography collection methods. Residents affected by changes to Barataria Bay have a continued need to document environmental damage, which Public Lab can help meet. For instance, Public Lab's aerial photography will be useful in determining the long-term effects that the Deepwater Horizon's oil has had on erosion, including following up on previous published research. While the aerial photography data collected by Public Lab methods have been sufficient to map shoreline change, data that have less than ten centimeters resolution offer an opportunity for greater analysis. For example, some of the Barataria survey data approach two centimeters resolution, which is the threshold for identifying birds. Finally, the mapping methods described here have been used to map the progress of land regeneration and loss and the health of important ecosystems in New York's Jamaica Bay (Wirth 2012), Pamet Marsh in Massachusetts (Ward 2013), and La Albufera in El Saler, Spain.

## SUMMARY

The rapid loss of wetlands in Louisiana has created a political landscape in which ecosystem services are institutionalized on an unprecedented scale. While a great deal of restoration has already occurred, much more is required in the face of rising sea levels. Given the institutional forces that have led to ecological crisis, developing new accountability measures and systems that can be placed in the hands of citizens impacted by the performance of restoration programs is necessary for the success of these programs. From explaining the Deepwater Horizon disaster in 2010, to land-loss monitoring at Barataria Bay in 2011 and 2013, to systematically mapping urban restoration sites in 2014, Public Lab aerial monitoring methods have provided a path to educating and empowering the public to engage in scientific assessment and the governance of restoration. Public Lab's methods have now been replicated outside of the Gulf Coast, and new applications are demonstrating the value of low-cost, open-source mapping methods as a mechanism for capturing snapshots of landscapes in flux. Beyond mapping tools, Public Lab continues to develop and implement new monitoring tools that will continue to aid citizen science and help democratize science.

# NINETEEN

# Reef Check California

## Scuba-Diving Citizen Scientists Monitor Rocky Reef Ecosystems

JAN FREIWALD and ANDREW BEAHRS

In 1999, the California State Legislature passed the Marine Life Protection Act (MLPA), which sought to protect and preserve the state's underwater heritage by establishing a network of protected sites. The act was guided by six principal goals, including protecting the natural diversity and abundance of marine life, rebuilding populations of marine organisms, and preserving unique marine habitats. In addition, the MLPA sought to manage Marine Protected Areas (MPAs) as a coherent network and—crucially—to ensure that they were managed adaptively and on the basis of "sound scientific guidelines" (California Fish and Game Code, sections 2850–2863).

After a lack of funding and an inadequate number of engaged stakeholders led to the repeated failure of initial MPA establishment, a private-public partnership called the Marine Life Protection Act Initiative (MLPA Initiative) was established in 2004 to design and implement a network of MPAs (Weible 2008, Kirlin et al. 2013). Given both the physical realities and political and social concerns, the complexities of California's MPA system were considerable from the beginning. Ranging along a coastline of more than a thousand miles, the protected areas vary widely in accessibility, urban access, and local environment. The fact that California's Oregon and Mexico borders both include MPAs—Pyramid River State Marine Conservation Area and Tijuana River Mouth State Marine Conservation Area, respectively—is suggestive of the diversity in socioeconomic circumstances and local marine environments across the MPA network. This points to the necessity of a regional approach when addressing the challenges of designing an MPA network capable of incorporating these complexities.

Furthermore, it was clear that the diverse MPA habitats were under a broad spectrum of pressures, from commercial and recreational fishing to water pollution to urban runoff. The

MPAs were therefore assigned three different levels of protection: State Marine Conservation Areas restrict certain types of commercial and/or recreational fishing; State Marine Parks—largely restricted to areas within San Francisco Bay—prohibit commercial extraction; and State Marine Reserves forbid all extraction other than permitted scientific collection.

Clearly, fulfilling MPLA's mandate to establish a strong scientific basis for MPA management would require an active monitoring program (California Department of Fish and Game 2008). Understanding the effects of MPAs on marine life—and the success of MPAs in creating "safe zones" that would help sustain reproductive populations—demanded strong and consistent monitoring so that sites could be compared with each other and compared with themselves over time. What's more, MPA sites would need to be compared with a set of comparable but unprotected sites, which could provide a reference (i.e., control) that changes in MPAs could be measured against.

In 2005, as the MLPA Initiative began the implementation of MPAs along California's Central Coast, the Reef Check Foundation established Reef Check California (RCCA), a program that monitors rocky reefs and kelp forests. These forests of giant kelp *(Macrocystis pyrifera)* and other algae provide an iconic coastal habitat with an immense diversity of fish and invertebrates (Foster and Schiel 1985, Carr and Reed 2016). Reef Check itself was founded in 1996 to monitor and protect the health of the world's coral reefs by training teams of volunteer scuba divers as citizen scientists to monitor reefs near their communities. This initiated the collection of valuable data and inspired grassroots conservation efforts. Since its inception, the Reef Check Foundation has trained over thirty thousand volunteers in over eighty nations. RCCA was designed to build upon the success of its international parent program, and a protocol was designed in collaboration with an existing rocky reef monitoring program in California, namely the subtidal monitoring program of the Partnership for Interdisciplinary Studies of Coastal Oceans (PISCO).

The creation of the RCCA program in 2005, in parallel to the MLPA Initiative's process of creating an MPA network, was intended to meet the need (evident in the MLPA legislation itself) for a long-term monitoring program. At the same time, the MLPA Initiative had committed to an open process of public participation in the implementation of California's MPA network (Sayce et al. 2013). Both developments created an environment in which a citizen science–driven, cost-effective, long-term monitoring program could make a substantial contribution to MPA implementation by establishing an initial ecological baseline and—vitally—offering cost-effective, long-term monitoring of MPAs into the future.

Reef Check thus designed its underwater monitoring program to maximize the usefulness of citizen scientist–collected data for MPA management. Key steps during program design strove to ensure data quality and longevity. For example, the monitoring program was modeled on an existing protocol for monitoring kelp forests; the targeted end-users of the data were engaged early on in program development; and strict procedures for data management and quality control were put in place from the beginning.

The model for the RCCA monitoring protocol was an academic monitoring program for kelp forests developed and run by PISCO, which helped ensure the compatibility of RCCA data with other monitoring data collected in California (Gillett et al. 2012). Through close collaboration with PISCO scientists, and several field tests over the first year of the program, the citizen science monitoring protocol was designed to produce data directly compatible with the academic monitoring program's data. For example, sampling units (in this case, a transect thirty meters long and two meters wide) were kept uniform, while taxonomic groupings and habitat types surveyed were both easily comparable across the academic and citizen science programs.

Intended end-users were engaged with protocol development throughout. Leading scientists and administrators from the California Department of Fish and Wildlife (CDFW) worked with academic scientists on protocol development, ensuring that future data would fill the department's needs. This collaboration resulted in a memorandum of understanding between the Reef Check Foundation and the CDFW, which recognized the scientific rigor of the RCCA's protocol and training methods as well as the usefulness of data and information provided by RCCA for the department's MLPA program.

Establishing the scientific rigor and usefulness of RCCA data for later management use during program development led to strict data-management and quality-assurance procedures. These procedures were implemented at key steps throughout the program, from the formal training volunteers receive before collecting data, to annual volunteer recertification, to strict data-quality protocols in both the field and during data entry and management. Detailed training protocols, hard-copy and online training materials, and the online data entry and display of RCCA's Global Reef Tracker system (http://data.reefcheck.org) are direct results of this early focus on scientific rigor and data use.

All of the aforementioned steps have helped RCCA make a large contribution to MPA monitoring and to the establishment of ecological baselines in all of California's MPLA regions. RCCA now maintains a network of some 250 volunteer divers dedicated to monitoring the rocky reefs off California's shores. By educating these dedicated members of the public about the marine environment and training them in monitoring techniques, the program provides consistent, robust data about life within the MPA network to resource managers throughout the state (Freiwald and Wehrenberg 2013, OST and CDFW 2013, Freiwald and Wisniewski 2015).

## REEF CHECK TRAINING AND MONITORING PROTOCOLS

Before being allowed to participate in RCCA's citizen scientist training, prospective volunteers must possess considerable experience in cold-water diving: a minimum of thirty lifetime dives is required, fifteen of which must be in California or other temperate (cooler than 65°F) waters, and six of which must have been during the preceding year. Volunteers also need to provide all their own dive gear. These requirements not only provide a vital safety margin through familiarity with equipment and the local marine environment, but also help ensure that volunteers are capable of the considerable task loading inherent to underwater surveys.

Qualified volunteers undergo four days of immersive training, including twelve hours of classroom time, a half-day pool scuba session, and six ocean dives over the course of two days. The classroom sessions focus on reef ecology, conservation and MPA science, and identification of seventy-three selected indicator species of algae (both native and invasive), fish, and invertebrates. A considerable amount of time is devoted to survey techniques specific to each of the three indicator-species categories and to uniform point contact (UPC) surveys that record the substrate, biological substrate cover (i.e., attached organisms), and rugosity of a rocky reef site (Freiwald et al. 2015).

Prior to certification as a Reef Check California diver, volunteers must demonstrate consistent proficiency in at least one of the four survey categories (fish, invertebrates, algae, and UPC). Proficiency is judged by both organism identification, with a score of more than 85 percent required during a written test, and underwater survey skills. Acquiring both can often prove challenging, and in spite of thorough training and the base of experience required, volunteers typically receive initial certification in only one to three of the survey types. This tiered approach to certification allows volunteers to collect data according to their skill level without compromising data quality for more technically challenging survey types.

Certification for fish surveys during the initial four-day training is especially rare, and almost exclusively limited to individuals with extensive, and often professional, previous experience identifying fish. For others, the challenge of counting, identifying, and sizing fish is complicated by additional gear and task loading, by the mobility of the species surveyed, and by light refraction that can cause even experienced scuba divers to badly misjudge size without a good deal of concentrated practice. This is usually too much to master during an initial training period, and several dedicated ocean sessions are typically required before a volunteer can demonstrate consistent proficiency in fish identification and sizing.

Each of the four survey types involves recording data during an individual pass along a thirty-meter-long transect line (figure 19.1). A survey begins when a buddy pair of divers descend to a depth of five to eighteen meters (for safety reasons, no sites deeper than eighteen meters are surveyed) and conduct a visibility check using the transect tape. Fish surveys require a minimum of three meters of visibility, which sometimes leads to surveys being postponed entirely even after full dive teams have entered a site.

If visibility permits, a fish-certified diver affixes the end of his transect to a kelp stipe or other stationary object, records the time, confirms a bearing consistent with the site's survey plan, and begins the fish survey. The diver looks for fish belonging to any of thirty-five indicator species within an imagined square two meters on a side, recording each by species and estimated length. Proceeding along the bottom for several meters, the diver plays out the transect line while using a light to check crevices and beneath overhangs for any fish missed during the initial scan. After stopping to record any observed fish, the diver then repeats the entire cycle until the end of the thirty-meter transect has been reached. To ensure relative uniformity of sample size, the fish survey is expected to take five to ten minutes from start to finish. Similarly, the transect must not cross more than ten contiguous meters of sandy

FIGURE 19.1. A Reef Check California volunteer swims along a transect tape counting organisms in a standardized area along the seafloor. Credit: photo by Jan Freiwald.

bottom or vary in depth by more than four meters, which ensures a minimum of consistency in both rocky bottom surveyed and variation in vertical zonation.

While the fish survey is under way, the second diver follows behind, conducting either the invertebrate or (less commonly) the algae survey. Both survey types include an area of one meter on either side of the transect, for a total surveyed area of sixty square meters. Invertebrate surveys require lights, which enable divers to survey the ubiquitous cracks and crevices present on most rocky reefs. The invertebrate surveys include twenty-eight indicator species, seven of which are abalone species. All these species have minimum size requirements to be included in organism counts (typically 2.5 cm; 10 cm is the minimum for gorgonians and sea anemones). In addition, any abalone found within the transect must be measured to the best of the diver's ability (if this is impossible due to the organism's orientation within a crack, the diver simply notes its presence without recording size data).

Several survey protocols help make invertebrate surveys manageable and consistent. Should a diver count fifty organisms of a single type—typically bat stars, purple sea urchins, or, in Northern California sites, red abalone—he or she records the distance along the transect at which the count was reached. Along the same lines, invertebrate surveys are expected to be completed within fifteen to twenty minutes, because a relatively consistent pace along the transect line by volunteers helps maintain consistency in data gathered.

Special procedures regarding sea urchin species, which represent one of the crucial ecological components of the sites discussed later in this chapter, are worth describing here. Three are RCCA indicator species: red urchin *(Strongylocentrotus franciscanus)*, purple urchin *(S. purpuratus)*, and crowned urchin *(Centrostephanus coronatus)*. In addition to counting all three species on normal invertebrate surveys, RCCA divers perform annual size frequency surveys for red and purple urchins at sites where urchins are abundant. Since urchins at these sites have already been surveyed on regular transects, no density data are recorded. Instead, divers use calipers to record the sizes of the first one hundred urchins they encounter, measuring each body—or test—to the nearest centimeter.

The final organism counts, of algae, are generally conducted by the same diver upon conclusion of the invertebrate survey. Returning along the same transect, the diver records the presence of five native and four invasive species. Most of the native species—southern sea palm, stalked kelp, bull kelp, and laminaria—consist of one stipe to the organism. However, the anchoring holdfast of a single giant kelp *(M. pyrifera)* can include scores of stipes, so in this instance volunteers are required to count all associated stipes.

Invasive algal species are an important exception to the normal practice of recording species within the sixty-square-meter transect. *Undaria pinnatifida, Caulerpa taxifolia,* and two *Sargassum* species are all recorded, and, if possible, photographed for later verification. While this presence/absence recording is also standard practice for Reef Check divers who observe rare species, such as giant sea bass, black abalone, and white abalone, the stakes are somewhat higher in the case of algae, given the tendency of invasive species to overrun entire sites or even larger areas (Marks et al. 2015).

The final survey type, the UPC, is intended to sample the physical structure of the transect being surveyed. A UPC survey consists of thirty records, each of which includes data on bottom substrate, cover, and rugosity. To perform the UPC survey, divers place their finger above a meter mark on the transect line, close their eyes, and blindly touch the bottom (this no-look touch helps negate the natural tendency to fall on more interesting substrate). The diver then proceeds to record *substrate*, consisting of four size categories from sand to cobble to boulder to reef; *cover*, which includes fleshy and coralline algae, sessile and mobile invertebrates, and seagrasses; and *rugosity*, or vertical relief, which is gauged by estimating the highest and the lowest point within a box (0.5 × 1 m) beginning at the UPC point. The diver repeats this procedure at every meter marker along the transect, thus providing a robust sample of the habitat being recorded.

As of 2016, RCCA surveys some ninety rocky reef sites throughout California (figure 19.2). A full survey of a single site includes eighteen transects: six "core transects," which include all four data categories described above; and twelve fish-only transects. By flanking each core transect with fish-only transects on either side, RCCA gathers more consistent data on these mobile, and thus inherently variable, species. Clearly, an RCCA survey is a substantial operation, with up to a dozen divers entering the water with clear instructions

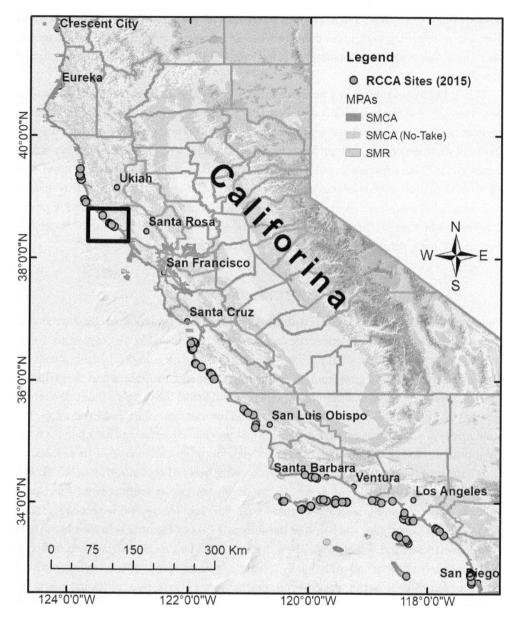

FIGURE 19.2. Map of Reef Check California (RCCA) monitoring sites. Since 2006, RCCA has built a network of about ninety monitoring sites in the state. Black frame indicates the Sonoma County sites discussed in this chapter. Legend abbreviations: MPAs = Marine Protected Areas; SMCA = State Marine Conservation Area; SMR = State Marine Reserve.

on start points, bearings, and surveys to be conducted. Although most surveys are completed in a single day, ocean conditions or a lack of personnel sometimes require a return visit (these follow-up surveys must take place within thirty days for a site to be considered complete).

While RCCA procedures and protocols are designed for consistent, accurate data collection, the profile of the typical RCCA volunteer is just as important. Among volunteers, 95 percent have a four-year college degree, and 48 percent have a graduate degree. Furthermore, all volunteers possess a high level of local diving experience. This combination of education and in-water experience, when coupled with the large investment in time, travel, and equipment maintenance necessary during an RCCA survey season, makes for a passionate and committed volunteer cohort. This passion is also reflected in the high retention rate of volunteers who have participated in RCCA surveys, with 70–80 percent of surveyors returning the following year to get recertified and many participating year after year.

## CASE STUDY: SONOMA COAST KELP FOREST

Four sites along the Sonoma County coast provide a useful case study of Reef Check's monitoring of California's kelp forests, one of the state's most important and iconic ecosystems. Growing on rocky reefs from the intertidal down to about twenty meters in depth, these forests are formed by large algae that grow from the seafloor up to the ocean surface, where they typically form a canopy of seaweed visible from shore (Foster and Schiel 1985). Below the surface, these forests provide habitat for many species of fish and invertebrates (Carr and Reed 2016). Not unlike trees in a forest on land, the fast-growing kelp provides structure and habitat in the water column above the seafloor as well as food and shelter for the species inhabiting these reefs.

There are two primary species of canopy-forming seaweed on California's reefs (Foster and Schiel 1985). Giant kelp *(M. pyrifera)* is the dominant species south of Point Año Nuevo (midway between San Francisco and Santa Cruz). Bull kelp *(Nereocystis luetkeana)* is found north of Point Conception and becomes the dominate canopy-forming kelp species north of Año Nuevo (Foster and Schiel 1985). These two species differ substantially in their morphology. Giant kelp are attached to the rocky substrate by large holdfasts from which multiple stipes (or fronds) grow to the surface. Each stipe has many pneumatocysts (floats), one at the base of every blade, that provide the lift for the stipes to reach the surface. Giant kelp is perennial, and individuals can live up to nine years, but in most cases large winter storms remove individuals before they can reach this age (Schiel and Foster 2006). Bull kelp, in contrast, consists of a single stipe ending in a large pneumatocyst from which multiple blades radiate (the main stipe resembles a heavy bullwhip in shape, giving the species its name). Bull kelp is an annual species, and mature plants can reach over twenty meters in length, with their floats and fronds forming a dense canopy at the ocean's surface (Springer et al. 2006). Reef Check monitors the density of both these kelp species at all of its sites.

All four locations discussed here (Fort Ross, Stillwater Cove, Ocean Cove, and Gerstle Cove) are small coves characterized by bedrock and boulders with relatively high relief (Frei-

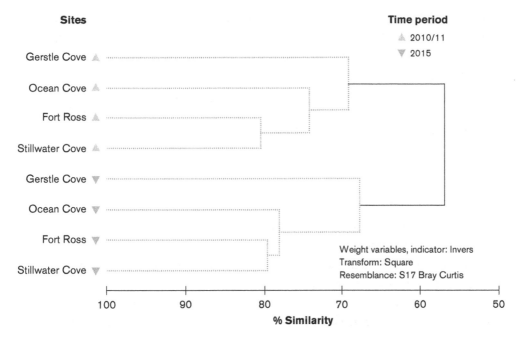

FIGURE 19.3. Cluster analysis of rocky reef community composition at four Reef Check California monitoring sites along the Sonoma County coast, indicating similarity among sites. Dotted branches indicate communities that are not significantly different from each other ($P = 0.05$). During the Marine Protected Area baseline monitoring period in 2010–11, all these sites were similar to each other. By 2015, community composition had changed at all four sites; all the sites were still similar to each other, but each differed significantly in comparison to its own community composition during the baseline period.

wald and Wehrenberg 2013). Reef Check volunteers have monitored bull kelp, other algae, fish, and invertebrates at all four sites since the beginning of the regional monitoring program in 2007. These four sites are all accessible from shore and relatively shallow, averaging between four and six-and-a-half meters. Divers typically monitor the reefs by swimming out from the beach during the months of July to October every year.

In 2010, new MPAs were established along this stretch of California's North Coast when the MPA network under development by the MLPA Initiative was expanded in what is referred to as the North Central Coast Region, which ranges from Pigeon Point (south of San Francisco) to Alder Creek (near Point Arena in Mendocino County). As in other MPA regions, RCCA became part of the consortium of monitoring programs that were working to establish a marine ecosystem baseline, which included characterizing the ecological communities of fish, invertebrates, and algae at these four sites in 2010 and 2011 (Freiwald and Wehrenberg 2013). The impression of overall similarity among the four sites during the baseline period was confirmed by a multivariate cluster analysis in PRIMER (Clarke and Gorley 2006) that compared the presence and abundance of species at the sites and indicated similar ecological communities of fish, invertebrates, and algae at all four sites during those two years (figure 19.3).

In 2010 and 2011, bull kelp was abundant at most of these sites. Another algal species, stalked kelp *(Pterygophora californica)*, also commonly formed dense stands of subcanopy below the main canopy of bull kelp. Invertebrate species that were present included giant spined star *(Pisaster giganteus)*, sunflower star *(Pycnopodia helianthoides)*, and red abalone *(Haliotis rufescens)*. The two sea urchin species typically found in Northern California, purple urchin and red urchin, were found at very low numbers in Reef Check surveys. Common fish species included blue rockfish *(Sebastes mystinus)* and several species of surf perch (Freiwald and Wehrenberg 2013). Overall, the abundances of kelp and invertebrate species at these sites during the two years of baseline monitoring were similar to what Reef Check volunteers had recorded since 2007 (figure 19.4).

After the baseline monitoring period, Reef Check citizen scientists continued to monitor these four sites annually. However, other monitoring programs that had contributed to the study of local MPA baselines were discontinued and therefore, for the following four years (2012–15), Reef Check was the only annual and comprehensive ecosystem monitoring program studying the rocky reefs and kelp forest of this section of coast. This continued monitoring of the four sites by a small team of volunteers is a prime example of how long-term monitoring by citizen scientists can detect and document changes to the environment that might otherwise have gone unnoticed.

In the four years following baseline monitoring, RCCA documented enormous changes in the kelp forest ecosystems. While several sea star species and canopy kelps have completely disappeared, sea urchins have increased over a hundredfold compared to their previous densities, and, seemingly as a consequence, understory algae have declined (figure 19.4). RCCA's continued annual monitoring lets us reconstruct the events that have led to this complete shift of the rocky reef ecosystem from kelp forests to urchin barrens in only three years at these sites.

Beginning in 2013, as the widely reported sea star wasting syndrome hit the North American west coast (Hewson et al. 2014), sea star populations were decimated at the four sites (figure 19.4). During the 2014 surveys, sunflower stars had completely disappeared from all four sites. At the same time, bull kelp began disappearing at the three sites where it had been abundant (Fort Ross, Stillwater Cove, and Ocean Cove), and it has not returned since (figure 19.4).

While it is not entirely clear why the canopy-forming kelp disappeared at the same time that the wasting syndrome hit the sea star populations at these sites, it is likely related to the abnormally warm water in those years (Bond et al. 2015). Regardless of the initial causes, consequent changes to the community are likely a result of this concurrence. Sunflower stars, which can grow up to a meter in diameter, are voracious predators of invertebrate species (figure 19.5; Carr and Reed 2016). They have been observed to preferentially feed on purple sea urchins over other urchin species, and some experiments have suggested that their presence might reduce feeding rates of purple urchins on kelp by almost 80 percent (Moitoza and Phillips 1979, Byrnes et al. 2006).

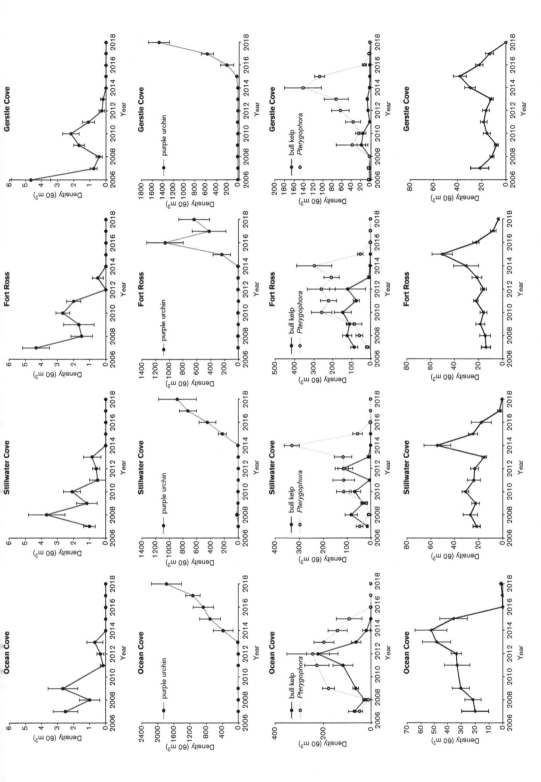

FIGURE 19.4. Trends in population densities of some important indicator species at four Reef Check California monitoring sites along the Sonoma County coast, 2007–2018: sunflower sea star (top row), purple sea urchin (second row), two algae (third row), and red abalone (bottom row).

FIGURE 19.5. Sunflower sea stars feeding on purple sea urchins at a Reef Check California monitoring site in Mendocino County. Credit: photo by Jan Freiwald.

Starting in 2014 at the Ocean Cove site, purple urchins began increasing in abundance, such that densities, which had previously been well below ten individuals per transect, increased to several hundred per transect (figure 19.4). The following year, purple urchins were present in similar densities at two of the other sites and continued to increase in density at Ocean Cove. Concurrently, densities of understory kelps, such as *Pterygophora*, showed declines (figures 19.4 and 19.6). Red abalone populations showed stable densities prior to 2013 and increased initially at all sites as bull kelp and sea stars disappeared, but in the following years (2016–18), red abalone decreased to levels so low that the recreational abalone fishery along California's North Coast was closed in 2017 by the California Fish and Game Commission (figure 19.4). Although fishing pressure did not cause this decline, the hope is that reducing additional mortality will help populations recover should environmental conditions change in the future. The initial increase in red abalone population densities observed during surveys in 2014 and 2015 may have been the result of behavioral changes (i.e., movement) in response to a lack of food. Red abalone, which mostly feed on drift algae, may have come out of the crevices where they usually reside in search of food and, therefore, may have been easier to count during those years. The consequent decline in their populations was caused by high mortality rates due to the lack of food. Reef Check's time-series data were presented to the Fish and Game Commission and helped inform their decision to close the fishery until populations recover.

Overall, the ecological communities at these four sites had changed dramatically between 2010–11 and 2015. By 2015, their ecological composition and structure had shifted, such that

FIGURE 19.6. Understory kelp *(Pterygophora)* at two Reef Check California monitoring sites in Northern California in 2012 (top) and 2015 (bottom). Credit: photo by Jan Freiwald.

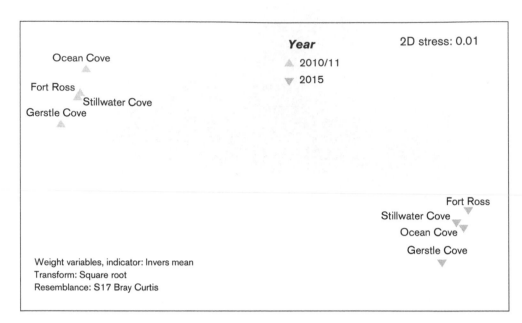

FIGURE 19.7. An NMDS (non-metric multidimensional scaling) plot generated in PRIMER (Clarke and Gorley 2006), showing the differences in community composition and structure at the four monitoring sites in 2010–11 and 2015. The closer together two sites are, the more similar their ecological communities. Sites are very similar to each other within each monitoring period, but the communities changed significantly between those periods.

communities were more similar to each other in that year than to themselves four years earlier (figure 19.7). The dramatic changes in kelp, sea star, and urchin populations have caused a community shift from kelp-dominated rocky reefs to urchin barrens, reducing similarity in the communities between the two study periods to less than 60 percent. The shift in the kelp forest community is likely due to multiple causes. Sea star wasting disease caused the decline of sea star populations (Hewson et al. 2014). Oceanographic conditions are likely to have played a role in the disappearance of bull kelp at the same time, by either reducing growth or leading to recruitment failures (Springer et al. 2006). The loss of canopy kelp and sea stars was followed by a sharp increase in purple urchin abundance, likely due to a recent increase in urchin recruitment as well as to behavioral changes leading to the emergence of purple urchins on the reef's surface, where they are actively feeding, rather than residing on the cracks and crevices where they previously fed on abundant drift algae (Foster 1990, Pearse 2006). The combination of recruitment and behavioral changes, induced by a lack of food and the disappearance of their main predator, could have led to the sudden emergence of urchins on these reefs, and their feeding is likely responsible for the subsequent reduction in understory kelps and the prevention of the return of bull kelp (Pearse and Hines 1987, Watanabe 1991).

Whatever the initial causes, Reef Check has witnessed a progression of events that appears to have led to a phase shift in these communities from kelp forests to urchin barrens

(Ling et al. 2009). Although understanding the mechanisms of these ecosystem changes will require more work, it is certain that without the long-term time-series data in this region provided by Reef Check citizen scientists, we would lack information to help clarify the chain of events that led to the differences in the ecosystem that we have seen since 2015 compared to the 2010–11 baseline and prior.

## SUMMARY

Reef Check California has built a large, sustainable monitoring program by training and guiding scuba-diving citizen scientists to survey the ocean ecosystems of California rocky reefs and kelp forests. Through this program, Reef Check has amassed one of the geographically largest and temporally longest ecological datasets focused on California's iconic kelp forest ecosystem. Rigorous training of volunteers and data quality control, continued collaboration with academic monitoring programs, and repeated comparison of RCCA's survey results with other monitoring outcomes have contributed to the success and long-term viability of its monitoring program. The initial involvement of agencies and academic scientists and the targeting of data users (i.e., CDFW) during program design were critical to successful implementation and to gaining the trust of stakeholders and the marine management community.

The high level of prerequisite skills and the detailed knowledge of the ecosystem and of scientific techniques gained during training select for a body of volunteers with educational backgrounds not unlike those of participants in academic monitoring programs. High levels of education and substantial initial investments of money (for dive gear and training) and time (for training and travel) lead to a high retention rate of volunteers who have successfully completed the training and participated in surveys. Many volunteers participate in the program over several years. On the other hand, these requirements also tend to select for a particular demographic and socioeconomic group of participants, somewhat limiting the educational value of the monitoring program. This limitation is addressed through some of the Reef Check Foundation's educational and outreach programs.

Participation in MPA baseline monitoring in all regions where MPAs were established is one of the greatest successes of RCCA and is unprecedented among citizen science programs in California. This involvement in the state-mandated monitoring of California's MPA network has allowed for the continued involvement of marine stakeholders past the initial, stakeholder-driven MPA implementation process. Through these baseline programs, RCCA has built volunteer and programmatic capacity for continued long-term monitoring of MPAs. This has led to Reef Check's continued monitoring of MPAs past the baseline periods, and in many cases the program has maintained the only subtidal monitoring, the other baseline participants having discontinued monitoring after the baseline periods ended.

Baseline and subsequent monitoring of Sonoma County sites by RCCA has provided a unique window into the rapid transformation of the communities at four local sites. Although such barrens have been common and sometimes persistent elsewhere in

California (Dayton et al. 1992, Graham 2004, Carr and Reed 2016), their relatively recent appearance in Sonoma County makes it possible to study their progression from the beginning.

Moreover, the appearance of such barrens along the North Coast has economic and cultural significance, with implications for the recreational abalone fishery. The closure of the recreational fishery has severe consequences for local communities, which depend on a seasonal influx of divers to support the local economy through lodging, restaurants, fishing supply stores, and other businesses. Many RCCA citizen scientists in Northern California are themselves abalone divers. Therefore, the data—and decisions based on the data—are directly relevant to their own lifestyles. Hopefully, this stakeholder involvement will continue to contribute to solid, science-based management decisions and help sustain this resource and this beloved activity when conditions change in the future.

# REFERENCES

## CHAPTER 1. WHAT IS CITIZEN SCIENCE?

Bonney, R., C. B. Cooper, J. L. Dickinson, S. Kelling, T. Phillips, K. V. Rosenberg, and J. Shirk. 2009. Citizen science: a developing tool for expanding science knowledge and scientific literacy. BioScience 59:977–984.

Dickinson, J. L., B. Zuckerberg, and D. N. Bonter. 2010. Citizen science as an ecological research tool: challenges and benefits. Annual Review of Ecology, Evolution, and Systematics 41:149–172.

Pocock, M. J. O., D. S. Chapman, L. J. Sheppard, and H. E. Roy. 2014. Choosing and using citizen science: a guide to when and how to use citizen science to monitor biodiversity and the environment. NERC/Centre for Ecology & Hydrology, Wallingford, UK. Available online: http://nora.nerc.ac.uk/id/eprint/510644/.

Shirk, J. L., H. L. Ballard, C. C. Wilderman, T. Phillips, A. Wiggins, R. Jordan, E. McCallie, M. Minarchek, B. V. Lewenstein, M. E. Krasny, and R. Bonney. 2012. Public participation in scientific research: a framework for deliberate design. Ecology and Society 17(2):29.

## CHAPTER 2. THE HISTORY OF CITIZEN SCIENCE IN ECOLOGY AND CONSERVATION

Bartomeus, I., J. S. Ascher, J. Gibbs, B. N. Danforth, D. L. Wagner, S. M. Hedtke, and R. Winfree. 2013. Historical changes in northeastern US bee pollinators related to shared ecological traits. Proceedings of the National Academy of Sciences USA 110:4656–4660.

Bartomeus, I., J. S. Ascher, D. Wagner, B. N. Danforth, S. Colla, S. Kornbluth, and R. Winfree. 2011. Climate-associated phenological advances in bee pollinators and bee-pollinated plants. Proceedings of the National Academy of Sciences USA 108:20645–20649.

Bled, F., J. Sauer, K. Pardieck, P. Doherty, and J. A. Royle. 2013. Modeling trends from North American Breeding Bird Survey data: a spatially explicit approach. PLoS ONE 8:e81867.

Bonney, R., H. Ballard, R. Jordan, E. McCallie, T. Phillips, J. Shirk, and C. C. Wilderman. 2009. Public participation in scientific research: defining the field and assessing its potential for informal science education. A CAISE Inquiry Group report. CAISE, Washington, DC.

Bonney, R., J. L. Shirk, T. B. Phillips, A. Wiggins, H. L. Ballard, A. J. Miller-Rushing, and J. K. Parrish. 2014. Next steps for citizen science. Science 343:1436–1437.

Brenna, B. 2011. Clergymen abiding in the fields: the making of the naturalist observer in eighteenth-century Norwegian natural history. Science in Context 24:143–166.

Brooks, S., A. Self, F. Toloni, and T. Sparks. 2014. Natural history museum collections provide information on phenological change in British butterflies since the late-nineteenth century. International Journal of Biometeorology 58:1749–1758.

Burkle, L.A., J.C. Marlin, and T.M. Knight. 2013. Plant-pollinator interactions over 120 years: loss of species, co-occurrence, and function. Science 339:1611–1615.

Butler, C.J. 2003. The disproportionate effect of global warming on the arrival dates of short-distance migratory birds in North America. Ibis 145:484–495.

Chandler, D.S., D. Manski, C. Donahue, and A. Alyokhin. 2012. Biodiversity of the Schoodic Peninsula: results of the insect and arachnid bioblitzes at the Schoodic District of Acadia National Park, Maine. Maine Agricultural and Forest Experiment Station, University of Maine, Orono, ME.

Chuine, I., P. Yiou, N. Viovy, B. Seguin, V. Daux, and E.L. Ladurie. 2004. Grape ripening as a past climate indicator. Nature 432:289–290.

Costello, M.J., R.M. May, and N.E. Stork. 2013. Can we name Earth's species before they go extinct? Science 339:413–416.

Dickinson, J.L., and R. Bonney, editors. 2012. Citizen Science: Public Participation in Environmental Research. Cornell University Press, Ithaca, NY.

Ellwood, E.R., R.B. Primack, and M.L. Talmadge. 2010. Effects of climate change on spring arrival times of birds in Thoreau's Concord from 1851 to 2007. The Condor 112:754–762.

Firehock, K., and J. West. 1995. A brief history of volunteer biological water monitoring using macroinvertebrates. Journal of the North American Benthological Society 14:197–202.

Granek, E.F., E.M.P. Madin, M.A. Brown, W. Figueira, D.S. Cameron, Z. Hogan, G. Kristianson, P. De Villiers, J.E. Williams, J. Post, S. Zahn, and R. Arlinghaus. 2008. Engaging recreational fishers in management and conservation: global case studies. Conservation Biology 22:1125–1134.

Greene, C.W., L.L. Gregory, G.H. Mittelhauser, S.C. Rooney, and J.E. Weber. 2005. Vascular flora of the Acadia National Park region, Maine. Rhodora 107:117–185.

Hampton, S.E., C.A. Strasser, J.J. Tewksbury, W.K. Gram, A.E. Budden, A.L. Batcheller, C.S. Duke, and J.H. Porter. 2013. Big data and the future of ecology. Frontiers in Ecology and the Environment 11:156–162.

Hopkins, A. 1918. Periodical events and natural law as guides to agricultural research and practice. Monthly Weather Review 9:1–42.

Jetz, W., J.M. McPherson, and R.P. Guralnick. 2012. Integrating biodiversity distribution knowledge: toward a global map of life. Trends in Ecology & Evolution 27:151–159.

Karl, T.R., J.M. Milillo, and T.C. Peterson. 2009. Global climate change impacts in the United States. U.S. Global Change Research Program, Cambridge, MA.

La Sorte, F.A., and W. Jetz. 2012. Tracking of climatic niche boundaries under recent climate change. Journal of Animal Ecology 81:914–925.

Lavoie, C., and A. Saint-Louis. 2008. Can a small park preserve its flora? A historical study of Bic National Park, Quebec. Botany 86:26–35.

Lavoie, C., A. Saint-Louis, G. Guay, E. Groeneveld, and P. Villeneuve. 2012. Naturalization of exotic plant species in north-eastern North America: trends and detection capacity. Diversity and Distributions 18:180–190.

Lepage, D., and C.M. Francis. 2002. Do feeder counts reliably indicate bird population changes? 21 years of winter bird counts in Ontario, Canada. The Condor 104:255–270.

Lowman, M., C. D'Avanzo, and C. Brewer. 2009. A national ecological network for research and education. Science 323:1172–1173.

McClenachan, L. 2009. Documenting loss of large trophy fish from the Florida Keys with historical photographs. Conservation Biology 23:636–643.

Miller-Rushing, A. J., D. W. Inouye, and R. B. Primack. 2008. How well do first flowering dates measure plant responses to climate change? The effects of population size and sampling frequency. Journal of Ecology 96:1289–1296.

Miller-Rushing, A. J., R. B. Primack, D. Primack, and S. Mukunda. 2006. Photographs and herbarium specimens as tools to document phenological changes in response to global warming. American Journal of Botany 93:1667–1674.

Moritz, C., J. L. Patton, C. J. Conroy, J. L. Parra, G. C. White, and S. R. Beissinger. 2008. Impact of a century of climate change on small-mammal communities in Yosemite National Park, USA. Science 322:261–264.

Pimm, S. L., C. N. Jenkins, R. Abell, T. M. Brooks, J. L. Gittleman, L. N. Joppa, P. H. Raven, C. M. Roberts, and J. O. Sexton. 2014. The biodiversity of species and their rates of extinction, distribution, and protection. Science 344.

Porter, R. 1978. Gentlemen and geology: the emergence of a scientific career, 1660–1920. The Historical Journal 21:809–836.

Primack, D., C. Imbres, R. B. Primack, A. J. Miller-Rushing, and P. Del Tredici. 2004. Herbarium specimens demonstrate earlier flowering times in response to warming in Boston. American Journal of Botany 91:1260–1264.

Primack, R. B., H. Higuchi, and A. J. Miller-Rushing. 2009. The impact of climate change on cherry trees and other species in Japan. Biological Conservation 142:1943–1949.

Primack, R. B., and A. J. Miller-Rushing. 2012. Uncovering, collecting, and analyzing records to investigate the ecological impacts of climate change: a template from Thoreau's Concord. BioScience 62:170–181.

Rosenberg, A. A., W. J. Bolster, K. E. Alexander, W. B. Leavenworth, A. B. Cooper, and M. G. McKenzie. 2005. The history of ocean resources: modeling cod biomass using historical records. Frontiers in Ecology and the Environment 3:84–90.

Scheper, J., M. Reemer, R. van Kats, W. A. Ozinga, G. T. J. van der Linden, J. H. J. Schaminée, H. Siepel, and D. Kleijn. 2014. Museum specimens reveal loss of pollen host plants as key factor driving wild bee decline in The Netherlands. Proceedings of the National Academy of Sciences USA 111:17552–17557.

Schwartz, M. D., J. L. Betancourt, and J. F. Weltzin. 2012. From Caprio's lilacs to the USA National Phenology Network. Frontiers in Ecology and the Environment 10:324–327.

Shirk, J. L., H. L. Ballard, C. C. Wilderman, T. Phillips, A. Wiggins, R. Jordan, E. McCallie, M. Minarchek, B. V. Lewenstein, M. E. Krasny, and R. Bonney. 2012. Public participation in scientific research: a framework for deliberate design. Ecology and Society 17(2):29.

Tian, H., L. C. Stige, B. Cazelles, K. L. Kausrud, R. Svarverud, N. C. Stenseth, and Z. Zhang. 2011. Reconstruction of a 1,910-y-long locust series reveals consistent associations with climate fluctuations in China. Proceedings of the National Academy of Sciences USA 108:14521–14526.

Tingley, M. W., W. B. Monahan, S. R. Beissinger, and C. Moritz. 2009. Birds track their Grinnellian niche through a century of climate change. Proceedings of the National Academy of Sciences USA 106:19637–19643.

Vetter, J. 2011a. Introduction: lay participation in the history of scientific observation. Science in Context 24:127–141.

Vetter, J. 2011b. Lay observers, telegraph lines, and Kansas weather: the field network as a mode of knowledge production. Science in Context 24:259–280.

Wals, A. E. J., M. Brody, J. Dillon, and R. B. Stevenson. 2014. Convergence between science and environmental education. Science 344:583–584.

Webb, R. H., D. E. Boyer, and R. M. Turner. 2010. Repeat Photography: Methods and Applications in the Natural Sciences. Island Press, Washington, DC.

Williams, E. S., M. W. Miller, T. J. Kreeger, R. H. Kahn, and E. T. Thorne. 2002. Chronic wasting disease of deer and elk: a review with recommendations for management. The Journal of Wildlife Management 66:551–563.

Willis, C. G., B. Ruhfel, R. B. Primack, A. J. Miller-Rushing, and C. C. Davis. 2008. Phylogenetic patterns of species loss in Thoreau's woods are driven by climate change. Proceedings of the National Academy of Sciences USA 105:17029–17033.

Willis, C. G., B. R. Ruhfel, R. B. Primack, A. J. Miller-Rushing, J. B. Losos, and C. C. Davis. 2010. Favorable climate change response explains non-native species' success in Thoreau's woods. PLoS ONE 5:e8878.

Zoellick, B., S. J. Nelson, and M. Schauffler. 2012. Participatory science and education: bringing both views into focus. Frontiers in Ecology and the Environment 10:310–313.

## CHAPTER 3. CURRENT APPROACHES TO CITIZEN SCIENCE

Bonney, R., T. B. Phillips, H. L. Ballard, and J. W. Enck. 2016. Can citizen science enhance public understanding of science? Public Understanding of Science 25:2–16.

Brofeldt, S., I. Theilade, N. D. Burgess, F. Danielsen, M. K. Poulsen, T. Adrian, T. N. Bang, A. Budiman, J. Jensen, A. E. Jensen, Y. Kurniawan, S. B. L. Lægaard, et al. 2014. Community monitoring of carbon stocks for REDD+: Does accuracy and cost change over time? Forests 5:1834–1854.

Chandler, M., L. See, H. Andrianandrasana, D. Becker, A. Berardi, R. Bodmer, P. de A. L. Constantino, J. Cousins, T. M. Crimmins, F. Danielsen, A. P. Giorgi, M. Huxham, et al. 2017. The value and opportunities of community- and citizen-based approaches to tropical forest biodiversity monitoring. Pages 223–281 in M. Gill, R. Jongman, S. Luque, B. Mora, M. Paganini, and Z. A. Szantoi (editors), Sourcebook of Methods and Procedures for Monitoring Essential Biodiversity Variables in Tropical Forests with Remote Sensing. GOFC—GOLD Land Cover Project Office, Wageningen University, The Netherlands.

Danielsen, F., N. D. Burgess, and A. Balmford. 2005. Monitoring matters: examining the potential of locally-based approaches. Biodiversity and Conservation 14:2507–2542.

Danielsen, F., P. M. Jensen, N. Burgess, R. Altamirano, P. A. Alviola, H. Andrianandrasana, J. S. Brashares, A. C. Burton, I. Coronado, N. Corpuz, M. Enghoff, J. Fjeldså, et al. 2014a. A multi-country assessment of tropical resource monitoring by local communities. BioScience 64:236–251.

Danielsen, F., N. Johnson, O. Lee, M. Fidel, L. Iversen, M. K. Poulsen, H. Eicken, A. Albin, S. G. Hansen, P. L. Pulsifer, P. Thorne, and M. Enghoff. 2020. Community-Based Monitoring in the Arctic. Fairbanks, USA: University of Alaska Press.

Danielsen, F., K. Pirhofer-Walzl, T. P. Adrian, D. R. Kapijimpanga, N. D. Burgess, P. M. Jensen, R. Bonney, M. Funder, A. Landa, N. Leverman, and J. Madsen. 2014b. Linking public participation in scientific research to the indicators and needs of international environmental agreements. Conservation Letters 7:12–24.

Fidel, M., N. Johnson, F. Danielsen, H. Eicken, L. Iversen, O. Lee, and C. Strawhacker. 2017. INTAROS Community-Based Monitoring Experience Exchange Workshop Report. Fairbanks, USA: Yukon River Inter-Tribal Watershed Council (YRITWC), University of Alaska Fairbanks, ELOKA, and Integrated Arctic Observing System Project (INTAROS).

Funder, M., F. Danielsen, Y. Ngaga, M. R. Nielsen, and M. K. Poulsen. 2013. Reshaping conservation: the social dynamics of participatory monitoring in Tanzania's community managed forests. Conservation and Society 11:218–232.

Johnson, N., C. Behe, F. Danielsen, E.-M. Krümmel, S. Nickels, and P. L. Pulsifer. 2016. Community-based monitoring and indigenous knowledge in a changing Arctic. A review for the Sustaining Arctic Observing Networks, Arctic Council. Final report to Sustaining Arctic Observing Networks, March. Inuit Circumpolar Council, Ottawa, ON.

Johnson, N., M. Fidel, F. Danielsen, L. Iversen, M. K. Poulsen, D. Hauser, and P. Pulsifer. 2018. INTAROS Community-Based Monitoring Experience Exchange Workshop Report. Québec City, Québec, Canada: ELOKA, Yukon River Inter-Tribal Watershed Council (YRITWC), University of Alaska Fairbanks, and Integrated Arctic Observing System Project (INTAROS).

Mustonen, T., and T. Tossavainen. 2018. Brook lampreys of life: towards holistic monitoring of boreal aquatic habitats using 'subtle signs' and oral histories. Reviews in Fish Biology and Fisheries 28:657–665.

Pocock, M. J. O., M. Chandler, R. Bonney, I. Thornhill, A. Albin, T. August, S. Bachman, P. M. J. Brown, D. G. F. Cunha, A. Grez, C. Jackson, M. Peters, et al. 2018. Chapter six—a vision for global biodiversity monitoring with citizen science. Advances in Ecological Research 59:169–223.

Shirk, J. L., H. L. Ballard, C. C. Wilderman, T. Phillips, A. Wiggins, R. Jordan, E. McCallie, M. Minarchek, B. V. Lewenstein, M. E. Krasny, and R. Bonney. 2012. Public participation in scientific research: a framework for deliberate design. Ecology and Society 17(2):29.

Støttrup, J. G., A. Kokkalis, E. J. Brown, J. Olsen, S. K. Andersen, and E. M. Pedersen. 2018. Harvesting geo-spatial data on fish assemblages through citizen science. Fisheries Research 208:86–96.

Tengö, M., R. Hill, P. Malmer, C. M. Raymond, M. Spierenburg, F. Danielsen, T. Elmqvist, and C. Folke. 2017. Weaving knowledge systems in IPBES, CBD and beyond—lessons learned for sustainability. Current Opinion in Environmental Sustainability 26: 17–25.

Venturelli, P. A., K. Hyder, and C. Skov. 2017. Angler apps as a source of recreational fisheries data: opportunities, challenges and proposed standards. Fish and Fisheries 18:578–595.

Wilderman, C. C., A. Barron, and L. Imgrund. 2004. Top down or bottom up? ALLARM's experience with two operational models for community science. Article 235 *in* Proceedings of the 4th National Monitoring Conference, May 17–20, 2004, Chattanooga, TN.

Zhao, M., S. Brofeldt, Q. Li, J. Xu, F. Danielsen, S. B. L. Læssøe, M. K. Poulsen, A. Gottlieb, J. F. Maxwell, and I. Theilade. 2016. Can community members identify tropical tree species for REDD+ carbon and biodiversity measurements? PLoS ONE 11:e0152061.

CHAPTER 4. PROJECT PLANNING AND DESIGN

August, T., M. Harvey, P. Lightfoot, D. Kilbey, T. Papadopoulos, and P. Jepson. 2015. Emerging technologies for biological recording. Biological Journal of the Linnean Society 115:731–749.

Ballard, H. L., L. D. Robinson, A. N. Young, G. B. Pauly, L. M. Higgins, R. F. Johnson, and J. C. Tweddle. 2017. Contributions to conservation outcomes by natural history museum-led citizen science: examining evidence and next steps. Biological Conservation 208:87–97.

Blaney, R. J. P., G. D. Jones, A. C. V. Philippe, and M. J. O. Pocock. 2016. Citizen science and environmental monitoring: towards a methodology for evaluating opportunities, costs and benefits. Final report on behalf of UK EOF. WRc, Fera Science, Centre for Ecology and Hydrology.

Bonney, R., H. Ballard, R. Jordan, E. McCallie, T. Phillips, J. Shirk, and C. C. Wilderman. 2009. Public participation in scientific research: defining the field and assessing its potential for informal science education. A CAISE Inquiry Group report. CAISE, Washington, DC.

Bowater, L., and K. Yeoman. 2012. Science Communication: A Practical Guide for Scientists. Wiley-Blackwell, London.

Bowser, A., A. Wiggins, and R. Stevenson. 2013. Data policies for public participation in scientific research: a primer. DataONE, Albuquerque, NM. www.birds.cornell.edu/citscitoolkit /toolkit/policy/Bowser%20et%20al%202013%2 0Data%20Policy%20Guide.pdf.

Ellis, R., and C. Waterton. 2005. Caught between the cartographic and the ethnographic imagination: the whereabouts of amateurs, professionals, and nature in knowing biodiversity. Environment and Planning D 23:673–693.

Geoghegan, H., A. Dyke, R. Pateman, S. West, and G. Everett. 2016. Understanding motivations for citizen science. Final report on behalf of UKEOF, University of Reading, Stockholm Environment Institute (University of York) and University of the West of England.

Grove-White, R., C. Waterton, R. Ellis, J. Vogel, G. Stevens, and B. Peacock. 2007. Amateurs as experts: harnessing new networks for biodiversity. Lancaster University, Lancaster, UK.

Hobbs, S. J., and P. C. L. White. 2012. Motivations and barriers in relation to community participation in biodiversity recording. Journal for Nature Conservation 20:364–373.

James, T. 2011a. Running a biological recording scheme or survey. National Biodiversity Network, UK. www.nbn.org.uk/Tools-Resources/NBN-Publications/Guidance- Documents.aspx.

James, T. 2011b. Improving wildlife data quality. National Biodiversity Network, UK. www.nbn .org.uk/Tools-Resources/NBN-Publications/Guidance-Documents.aspx.

Kapos, V., A. Balmford, R. Aveling, P. Bubb, P. Carey, A. Entwistle, J. Hopkins, T. Mulliken, R. Safford, A. Stattersfield, M. Walpole, and A. Manica. 2008. Calibrating conservation: new tools for measuring success. Conservation Letters 1:155–164.

McKinley, D. C., A. J. Miller-Rushing, H. L. Ballard, R. Bonney, H. Brown, S. Cook-Patton, D. M. Evans, R. A. French, J. K. Parrish, T. B. Phillips, S. F. Ryan, L. A. Shanley, et al. 2017. Citizen science can improve conservation science, natural resource management, and environmental protection. Biological Conservation 208:15–28.

Pandya, R. E. 2012. A framework for engaging diverse communities in citizen science in the US. Frontiers in Ecology and the Environment 10:314–317.

Phillips, T. B., M. Ferguson, M. Minarchek, N. Porticella, and R. Bonney. 2014. User's Guide for Evaluating Learning Outcomes in Citizen Science. Cornell Lab of Ornithology, Ithaca, NY.

Pocock, M. J. O., M. Chandler, R. Bonney, I. Thornhill, A. Albin, T. August, S. Bachman, P. M. J. Brown, D. G. F. Cunha, A. Grez, C. Jackson, M. Peters, et al. 2018. Chapter six—a vision for global biodiversity monitoring with citizen science. Advances in Ecological Research 59:169–223.

Pocock, M. J. O., D. S. Chapman, L. J. Sheppard, and H. E. Roy. 2014. Choosing and using citizen science: a guide to when and how to use citizen science to monitor biodiversity and the environment. NERC/Centre for Ecology & Hydrology, Wallingford, UK. Available online: http://nora.nerc.ac.uk/id/eprint/510644/.

Pocock, M. J. O, J. C. Tweddle, J. Savage, L. D. Robinson, and H. E. Roy. 2017. The diversity and evolution of ecological and environmental citizen science. PLoS ONE 12:e0172579.

Prysby, M., and P. Super, editors. 2007. Director's Guide to Best Practices: Programming—Citizen Science. Association of Nature Center Administrators, Logan, UT.

Ramirez-Andreotta, M. D., M. L. Brusseau, J. F. Artiola, R. M. Maier, and A. J. Gandolfi. 2014. Environmental research translation: enhancing interactions with communities at contaminated sites. Science of the Total Environment 497–498:651–654.

RCUK. 2011. Evaluation: Practical Guidelines. Research Councils UK. www.ukri.org/files/legacy /publications/evaluationguide-pdf/.

Resnik, D. B., K. C. Elliott, and A. K. Miller. 2015. A framework for addressing ethical issues in citizen science. Environmental Science & Policy 54:475–481.

Robinson, L. D., J. L. Cawthray, S. E. West, A. Bonn, and J. Ansine. 2018. Ten principles of citizen science. Pages 27–40 *in* S. Hecker, M. Haklay, A. Bowser, Z. Makuch, J. Vogel, and A. Bonn (editors), Citizen Science: Innovation in Open Science, Society and Policy. UCL Press, London.

Robinson, L. D., J. C. Tweddle, M. C. Postles, S. E. West, and J. Sewell. 2013. Guide to running a BioBlitz. Natural History Museum, Bristol Natural History Consortium, University of York and Marine Biological Association, UK. www.bnhc.org.uk/wp-content/uploads /2014/04/BioBlitz-Guide-2013.pdf.

Roy, H. E., M. J. O. Pocock, C. D. Preston, D. B. Roy, J. Savage, J. C. Tweddle, and L. D. Robinson. 2012. Understanding citizen science & environmental monitoring. Final report on behalf of UK-EOF. NERC Centre for Ecology & Hydrology and Natural History Museum, UK. www.ukeof.org.uk/resources/citizen-science-resources/.

Scassa, T., and H. Chung. 2015. Managing intellectual property rights in citizen science: a guide for researchers and citizen scientists. Woodrow Wilson International Center for Scholars, Washington, DC. www.wilsoncenter.org/sites/default/files/managing_intellectual_ property_rights_citizen_science_scassa_chung.pdf.

Shirk, J. L., H. L. Ballard, C. C. Wilderman, T. Phillips, A. Wiggins, R. Jordan, E. McCallie, M. Minarchek, B. V. Lewenstein, M. E. Krasny, and R. Bonney. 2012. Public participation in scientific research: a framework for deliberate design. Ecology and Society 17(2):29.

Tweddle, J. C., L. D. Robinson, M. J. O. Pocock, and H. E. Roy. 2012. Guide to citizen science: developing, implementing and evaluating citizen science to study biodiversity and the environment in the UK. Natural History Museum and NERC Centre for Ecology and Hydrology for UK-EOF. www.ukeof.org.uk/resources/citizen-science-resources/.

Wiggins, A., R. Bonney, E. Graham, S. Henderson, S. Kelling, R. Littauer, G. LeBuhn, K. Lotts, W. Michener, G. Newman, E. Russell, R. Stevenson, and J. Weltzin. 2013. Data management guide for public participation in scientific research. DataONE, Albuquerque, NM. www.dataone.org/sites/all/documents/DataONE-PPSR-DataManagementGuide .pdf.

## CHAPTER 5. LEGAL, ETHICAL, AND POLICY CONSIDERATIONS

Bleizeffer, D. 2016. Court will hear case against data trespass laws. Lake Tahoe News.

Bowser, A., and A. Wiggins. 2016. Privacy in participatory research: advancing policy to support human computation. Human Computation 2.1:19–44.

Bowser, A., A. Wiggins, and R. Stevenson. 2013. Data policies for public participation in scientific research: a primer. DataONE, Albuquerque, NM. www.birds.cornell.edu/citscitoolkit /toolkit/policy/Bowser%20et%20al%202013%2 0Data%20Policy%20Guide.pdf.

Cotten, D., and M. Cotton. 1997. Legal Aspects of Waivers in Sports, Recreation, and Fitness. PRC, Canton, OH.

El Emam, K. 2010. Risk-based de-identification of health data. IEEE Security Privacy 8:64–67.

Evans, D. M., J. P. Che-Castaldo, D. Crouse, F. W. Davis, R. Epanchin-Niell, C. H. Flather, R. K. Frohlich, D. D. Goble, Y.-W. Li, T. D. Male, L. L. Master, M. P. Moskwik, et al. 2016. Species recovery in the United States: increasing the effectiveness of the endangered species act. Issues in Ecology 20:1–29.

Gellman, R. 2015. Crowdsourcing, citizen science, and the law: legal issues impacting federal agencies. Woodrow Wilson International Center for Scholars, Washington, DC.

McElfish, J., J. Pendergrass, and T. Fox. 2016. Clearing the path: citizen science and public decision making in the United States. Woodrow Wilson International Center for Scholars, Washington, DC.

National Commission for the Protection of Human Subjects of Biomedical and Behavioral Research. 1978. The Belmont Report. Retrieved from http://videocast.nih.gov/pdf/ohrp_appendix_belmont_report_vol_2.pdf.

National Science Foundation. 2019. Proposal and award policies and procedures guide. www.nsf.gov/funding/pgm_summ.jsp.

Organ, S., and M. Corcoran. 2008. Your web site's "terms of use": are they enforceable? Privacy & Data Security Law Journal 3:110–114.

Robson, E. 2012. Responding to liability: evaluating and reducing tort liability for digital volunteers. Woodrow Wilson International Center for Scholars, Washington, DC.

Scassa, T., and H. Chung. 2015. Managing intellectual property rights in citizen science: a guide for researchers and citizen scientists. Woodrow Wilson International Center for Scholars, Washington, DC. www.wilsoncenter.org/sites/default/files/managing_intellectual_property_rights_citizen_science_scassa_chung.pdf.

Smith, B. 2014. Agency liability stemming from citizen-generated data. Woodrow Wilson International Center for Scholars, Washington, DC.

U.S. Department of Health and Human Services. 2015. Guidance regarding mobile methods for deidentification of protected health information in accordance with the Health Insurance Portability and Accountability Act (HIPPA) Privacy Rule. www.hhs.gov/ocr/privacy/hipaa/understanding/coveredentities/De-identification/guidance.html#_edn1.

CHAPTER 6. RECRUITMENT AND BUILDING THE TEAM

Allen, J., and R. Guralnick. 2014. How long and when do you transcribe records in Notes from Nature, and other neat ways to look at your (amazing) effort. https://blog.notesfromnature.org/2014/07/28/how-long-and-when-do-you-transcribe-records-in-notes-from-nature-and-other-neat-ways-to-look-at-your-amazing-effort/.

Appalachian Trail Conservancy. 2019. Natural and cultural resource management. https://appalachiantrail.org/home/conservation/stewardship.

Asah, S. T., and D. J. Blahna. 2012. Motivational functionalism and urban conservation stewardship: implications for volunteer involvement. Conservation Letters 5: 470–477.

Bodin, Ö., B. Crona, and H. Ernstson. 2006. Social networks in natural resource management: what is there to learn from a structural perspective? Ecology and Society 11(2):r2.

Bruyere, B., and S. Rappe. 2007. Identifying the motivations of environmental volunteers. Journal of Environmental Planning and Management 50:503–516.

Clary, E. G., M. Snyder, and R. Ridge. 1992. Volunteers' motivations: a functional strategy for the recruitment, placement, and retention of volunteers. Nonprofit Management and Leadership 2:333–350.

Clery, D. 2011. Galaxy Zoo volunteers share pain and glory of research. Science 333:173–175.

EarthWatch Institute. 2014. EarthWatch Fast Facts. https://earthwatch.org/about/fast-facts.

Empire State College. 2016. Citizen science central: the beetle project. www.esc.edu/citizen-science/.

Eveleigh, A., C. Jennett, A. Blanford, P. Brohan, and A. Cox. 2014. Designing for dabblers and deterring drop-outs in citizen science. Pages 2985–2994 *in* Proceedings of the SIGCHI Conference on Human Factors in Computing Systems.

Freeman, R. B. 1997. Working for nothing: the supply of volunteer labor. Journal of Labor Economics 15(1) Part 2:S140–S166.

Grese, R. E., R. Kaplan, R. L. Ryan, and J. Buxton. 2001. Psychological benefits of volunteering in stewardship programs. Pages 265–280 *in* P. H. Gobster and R. B. Hull (editors), Restoring

Nature: Perspectives from the Social Sciences and Humanities. Island Press, Washington, DC.

Hazzah, L., S. Dolrenry, L. Naughton, C.T.T. Edwards, O. Mwebi, F. Kearney, and L. Frank. 2014. Efficacy of two lion conservation programs in Maasailand, Kenya. Conservation Biology 28:851–860.

Houle, B.J., B.J. Sagarin, and M.F. Kaplan. 2005. A functional approach to volunteerism: do volunteer motives predict task preference? Basic and Applied Social Psychology 27:337–344.

Jacobson, S.K., J.S. Carlton, and M.C. Monroe. 2012. Motivation and satisfaction of volunteers at a Florida natural resource agency. Journal of Park and Recreation Administration 30:51–67.

Kellogg Foundation. 2004. Logic model development guide. Kellogg Foundation, Battle Creek, MI.

Larese-Casanova, M., and M. Prysby. 2018. Engaging people in nature stewardship through Master Naturalist programs. Journal of Human-Wildlife Interactions 12:259–271.

Leung, M.W., I.H. Yen, and M. Minkler. 2004. Community based participatory research: a promising approach for increasing epidemiology's relevance in the 21st century. International Journal of Epidemiology 33:499–506.

McDougle, L.M., I. Greenspan, and F. Handy. 2011. Generation green: understanding the motivations and mechanisms influencing young adults' environmental volunteering. International Journal of Nonprofit and Voluntary Sector Marketing 16:325–341.

National Oceanic and Atmospheric Administration. 2019. Cooperative Observer Program. www .weather.gov/coop/Overview.

Nov, O., O. Arazy, and D. Anderson. 2011. Technology-mediated citizen science participation: a motivational model. Pages 249–256 in N. Nicolov and J.G. Shanahan (editors), Proceedings of the Fifth International AAAI Conference on Weblogs and Social Media. AAAI Press, Menlo Park, CA.

Porticella, N., S. Bonfield, T. DeFalco, A. Fumarolo, C. Garibay, E. Jolly, L.H. Migus, R. Pandya, K. Purcell, J. Rowden, F. Stevenson, and A. Switzer. 2013. Promising practices for community partnerships: a call to support more inclusive approaches to public participation in scientific research. Report commissioned by the Association of Science-Technology Centers, Washington, DC.

Raddick, M.J., G. Bracey, P.L. Gay, C.J. Lintott, C. Cardamone, P. Murray, K. Schawinski, A.S. Szalay, and J. Vandenberg. 2013. Galaxy Zoo: motivations of citizen scientists. Astronomy Education Review 12.

Reed, J., M.J. Raddick, A. Lardner, and K. Carney. 2013. An exploratory factor analysis of motivations for participating in Zooniverse, a collection of virtual citizen science projects. Proceedings of the 46th Hawaii International Conference on System Sciences (HICSS-46), Grand Wailea, HI, January, 2013.

Robinson, D.A., Jr., W.E. Jensen, and R.D. Applegate. 2000. Observer effect on a rural mail carrier survey population index. Wildlife Society Bulletin 28:330–332.

Robinson, J.C. 2008. Birding for Everyone: Encouraging People of Color to Become Birdwatchers. Wings-on-Disk, Marysville, OH.

Rotman, D., J. Preece, J. Hammock, K. Procita, D. Hansen, C. Parr, D. Lewis, and D. Jacobs. 2012. Pages 217–226 in Proceedings of the ACM 2012 Conference on Computer Supported Cooperative Work, Seattle, WA.

Ryan, R.L., R. Kaplan, and R.E. Grese. 2001. Predicting volunteer commitment in environmental stewardship programmes. Journal of Environmental Planning and Management 44:629–648.

Stepenuck, K.F., L.G. Woodson, B.W. Liukkonen, J.M. Iles, and T.S. Grant. 2010. Volunteer monitoring of *E. coli* in streams of the upper Midwestern United States: a comparison of methods. Environmental Monitoring and Assessment 174:625–633.

Swanson, A. B. 2014. Living with lions: spatiotemporal aspects of coexistence in savanna carni-
    vores. Doctoral dissertation, University of Minnesota.

Taylor, D. E. 2014. The state of diversity in environmental organizations. Report for Green 2.0,
    Washington, DC.

University of Minnesota Lion Project. 2014. Snapshot Serengeti. www.zooniverse.org/projects
    /zooniverse/snapshot-serengeti.

USA National Phenology Network. 2012. The Phenology Trail Guide: an experiential education
    tool for site-based community engagement. USA-NPN Education & Engagement Series
    2012–001.

USA National Phenology Network. 2015. Tucson Phenology Trail. www.usanpn.org/node
    /20430.

U.S. Geological Survey. 2019. Citizen science light trapping in Grand Canyon. www.usgs.gov
    /centers/sbsc/science/citizen-science-light-trapping-grand-canyon?qt-science_center_
    objects = 0#qt-science_center_objects.

Wachob, D. O., and B. Smith. 2007. Case study: a study of elk migration through a suburban/
    exurban landscape or monitoring elk migration in your neighborhood. Pages 45–51 *in*
    M. Prysby and P. Super (editors), Director's Guide to Best Practices: Programming Citizen
    Science. Association of Nature Center Administrators, Logan, UT.

West, S., and R. Pateman. 2016. Recruiting and retaining participants in citizen science: what can
    be learned from the volunteering literature? Citizen Science: Theory and Practice 1(2):15.
    http://doi.org/10.5334/cstp.8.

CHAPTER 7. RETAINING CITIZEN SCIENTISTS

Bernardo, H. L., P. Vitt, R. Goad, S. Masi, and T. M. Knight. 2018. Count population viability
    analysis finds that interacting local and regional threats affect the viability of a rare plant.
    Ecological Indicators 93:822–829.

Bird Conservation Network. 2019. www.bcnbirds.org/.

Chicago Botanic Garden. 2019. Plants of concern. www.plantsofconcern.org/.

Chicago Region Biodiversity Council. 1999. Chicago Wilderness biodiversity recovery plan.
    Chicago Region Biodiversity Council, Chicago, IL.

Chicago Wilderness. 2019. www.chicagowilderness.org/.

Chu, M., P. Leonard, and F. Stevenson. 2012. Growing the base for citizen science: recruiting and
    engaging participants. Pages 69–81 *in* J. L. Dickinson and R. Bonney (editors), Citizen
    Science: Public Participation in Environmental Research. Cornell University Press, Ithaca,
    NY.

Citizen Science Association. 2019. www.citizenscience.org/.

Cornell Lab of Ornithology. 2019. Citizen science central. www.birds.cornell.edu/citscitoolkit.

Echinacea Project. 2019. http://echinaceaproject.org/.

Fant, J. B., R. M. Holmstrom, E. Sirkin, J. R. Etterson, and S. Masi. 2008. Genetic structure
    of threatened native populations and propagules used for restoration in a clonal
    species, American beachgrass (*Ammophila breviligulata* Fern.). Restoration Ecology
    16:594–603.

Illinois Butterfly Monitoring Network. 2019. https://bfly.org/.

Kelling, S. 2012. Using bioinformatics in citizen science. Pages 58–68 *in* J. L. Dickinson and
    R. Bonney (editors), Citizen Science: Public Participation in Environmental Research.
    Cornell University Press, Ithaca, NY.

Latteier, C. 2001. Linking citizens with scientists: how Illinois EcoWatch uses volunteers to collect
    ecological data. Conservation in Practice 2:31–36.

Marinelli, J. 2007. The orchid keepers. Audubon Magazine, May/June.

Peggy Notebaert Nature Museum. 2019. The Calling Frog Survey. http://frogsurvey.org/.

Ross, L. M. 1997. The Chicago Wilderness and its critics I. The Chicago Wilderness: a coalition for urban conservation. Restoration & Management Notes 15:17–24.

Sullivan, B. L., J. L. Aycrigg, J. H. Barry, R. E. Bonney, N. Bruns, C. B. Cooper, T. Damoulas, A. A. Dhondt, T. Dietterich, A. Farnsworth, D. Fink, J. W. Fitzpatrick, et al. 2014. The eBird enterprise: an integrated approach to development and application of citizen science. Biological Conservation 169:31–40.

UW Extension. 2019. Wisconsin's Citizen-Based Water Monitoring Network. http://watermonitoring.uwex.edu.

Vitt, P., K. Havens, B. E. Kendall, and T. M. Knight. 2009. Effects of community-level grassland management on the non-target rare annual *Agalinis auriculata*. Biological Conservation 142:798–805.

Wilson, E. O. 1984. Biophilia. Harvard University Press, Cambridge, MA.

## CHAPTER 8. TRAINING

Ballard, H. L., C. G. Dixon, and E. M. Harris. 2017. Youth-focused citizen science: examining the role of environmental science learning and agency for conservation. Biological Conservation 208:65–75.

Brooks, J. G., and M. G. Brooks. 1993. In search of understanding: the case for constructivist classrooms. Association for Supervision and Curriculum Development, Alexandria, VA.

Collins, A., J. S. Brown, and S. E. Newman. 1989. Cognitive apprenticeship: teaching the crafts of reading, writing, and mathematics. Pages 453–494 *in* L. B. Resnick (editor), Knowing, Learning, and Instruction: Essays in Honor of Robert Glaser. Lawrence Erlbaum Associates, Hillsdale, NJ.

Harris, E. M., and H. L. Ballard. 2018. Real science in the palm of your hand. Science and Children 55:31–37.

Haywood, B. K. 2014. A "sense of place" in public participation in scientific research. Science Education 98:64–83.

Jacobson, S. K., M. D. McDuff, and M. C. Monroe. 2016. Conservation Education and Outreach Techniques. Oxford University Press, Oxford, UK.

Jiguet, F. 2009. Method learning caused a first-time observer effect in a newly started breeding bird survey. Bird Study 56:253–258.

Jordan, R., J. G. Eherfeld, S. A. Gray, W. R. Brooks, D. V. How, and C. E. Hmelo-Sliver. 2012. Cognitive considerations in the development of citizen science projects. Pages 167–178 *in* J. L. Dickinson and R. Bonney (editors), Citizen Science: Public Participation in Environmental Research. Cornell University Press, Ithaca, NY.

Maslow, A. H. 1954. Motivation and Personality. Harper, New York.

Monroe, M. C. 2003. Two avenues for encouraging conservation behaviors. Human Ecology Review 10:113–125.

Sauer, J. R., B. G. Peterjohn, and W. A. Link. 1994. Observer differences in the North American Breeding Bird Survey. The Auk 111:50–62.

Schraw, G., K. J. Crippen, and K. Hartley. 2006. Promoting self-regulation in science education: Metacognition as part of a broader perspective on learning. Research in Science Education 36:111–139.

Wals, A. E. J., and J. Dillon. 2013. Conventional and emerging learning theories: implications and choices for educational researchers with a planetary consciousness. Pages 253–261 *in*

R. B. Stevenson, M. Brody, J. Dillon, and A. E. J. Wals (editors), International Handbook of Research on Environmental Education. Routledge, New York.

Wiggins, G. P., J. McTighe, L. J. Kiernan, and F. Frost. 1998. Understanding by design. Association for Supervision and Curriculum Development, Alexandria, VA.

## CHAPTER 9. COLLECTING HIGH-QUALITY DATA

Bonter, D. N., and C. B. Cooper. 2012. Data validation in citizen science: a case study from Project FeederWatch. Frontiers in Ecology and the Environment 10:305–307.

Conrad, C. C., and K. G. Hilchey. 2011. A review of citizen science and community-based environmental monitoring: issues and opportunities. Environmental Monitoring and Assessment 176:273–291.

Crall, A. W., G. J. Newman, C. S. Jarnevich, T. J. Stohlgren, D. M. Waller, and J. Graham. 2010. Improving and integrating data on invasive species collected by citizen scientists. Biological Invasions 12:3419–3428.

Crall, A. W., M. Renz, B. J. Panke, G. J. Newman, C. Chapin, J. Graham, and C. Bargeron. 2012. Developing cost-effective early detection networks for regional invasions. Biological Invasions 14:2461–2469.

Delaney, D. G., C. D. Sperling, C. S. Adams, and B. Leung. 2008. Marine invasive species: validation of citizen science and implications for national monitoring networks. Biological Invasions 10:117–128.

Dytham, C. 1999. Choosing and Using Statistics: A Biologist's Guide. Blackwell Science, Malden, MA.

Foster-Smith, J., and S. M. Evans. 2003. The value of marine ecological data collected by volunteers. Biological Conservation 113:199–213.

Genet, K. S., and L. G. Sargent. 2003. Evaluation of methods and data quality from a volunteer based amphibian call survey. Wildlife Society Bulletin 31:703–714.

Lewandowski, E., and H. Specht. 2015. Influence of volunteer and project characteristics on data quality of biological surveys. Conservation Biology 29:713–723.

Lindenmayer, D. B., and G. E. Likens. 2010. Effective Ecological Monitoring. CSIRO, Collingwood, Australia.

McComb, B., B. Zuckerberg, D. Vesely, and C. Jordan. 2010. Monitoring Animal Populations and Their Habitats: A Practitioner's Guide. CRC Press, New York.

Nature Editorial. 2015. Rise of the citizen scientist. Nature 524:265.

Sokal, R. R., and F. J. Rohlf. 1995. Biometry: The Principles and Practice of Statistics in Biological Research, 3rd ed. W. H. Freeman, New York.

Thornton, T., and J. Leahy. 2012. Trust in citizen science research: a case study of the Groundwater Education Through Water Evaluation and Testing program. Journal of the American Water Resources Association 48:1032–1040.

## CHAPTER 10. DATA MANAGEMENT AND VISUALIZATION

Bonney, R., H. Ballard, R. Jordan, E. McCallie, T. Phillips, J. Shirk, and C. C. Wilderman. 2009. Public participation in scientific research: defining the field and assessing its potential for informal science education. A CAISE Inquiry Group report. CAISE, Washington, DC.

DataONE. 2019a. Data lifecycle best practices. www.dataone.org/best-practices.

DataONE. 2019b. Software tool catalogue. www.dataone.org/all-software-tools.

Goring, S. J., K. C. Weathers, W. K. Dodds, P. A. Soranno, L. C. Sweet, K. S. Cheruvelil, J. S. Kominoski, J. Rüegg, A. M. Thorn, and R. M. Utz. 2014. Improving the culture of interdisciplinary collaboration in ecology by expanding measures of success. Frontiers in Ecology and the Environment 12:39–47.

Graham, J., G. Newman, C. Jarnevich, R. Shory, and T. J. Stohlgren. 2007. A global organism detection and monitoring system for non-native species. Ecological Informatics 2:177–183.

Greenland, S., and K. O'Rourke. 2008. Meta-analysis. Pages 652–682 in K. J. Rothman, S. Greenland, and T. L. Lash (editors), Modern Epidemiology, 3rd ed. Lippincott Williams & Wilkins, Philadelphia, PA.

Hampton, S. E., C. A. Strasser, J. J. Tewksbury, W. K. Gram, A. E. Budden, A. L. Batcheller, C. S. Duke, and J. H. Porter. 2013. Big data and the future of ecology. Frontiers in Ecology and the Environment 11:156–162.

Kelling, S., W. M. Hochachka, D. Fink, M. Riedewald, R. Caruana, G. Ballard, and G. Hooker. 2009. Data-intensive science: a new paradigm for biodiversity studies. BioScience 59:613–620.

Michener, W. K., J. W. Brunt, J. J. Helly, T. B. Kirchner, and S. G. Stafford. 1997. Nongeospatial metadata for the ecological sciences. Ecological Applications 7:330–342.

Michener, W. K., and M. Jones. 2011. Ecoinformatics: supporting ecology as a data-intensive science. Trends in Ecology & Evolution 27:83–93.

Nelson, G. L., and B. M. Graves. 2004. Anuran population monitoring: comparison of the North American Amphibian Monitoring Program's calling index with mark-recapture estimates for *Rana clamitans*. Journal of Herpetology 38:355–359.

Newman, G., J. Graham, A. Crall, and M. Laituri. 2011. The art and science of multi-scale citizen science support. Ecological Informatics 6:217–227.

Nielsen, M. 2012. Reinventing Discovery: The New Era of Networked Science. Princeton University Press, Princeton, NJ.

Rüegg, J., C. Gries, B. Bond-Lamberty, G. J. Bowen, B. S. Felzer, N. E. McIntyre, P. A. Soranno, K. L. Vanderbilt, and K. C. Weathers. 2014. Completing the data life cycle: using information management in macrosystems ecology research. Frontiers in Ecology and the Environment 12:24–30.

Salkind, N. J. 2011. Statistics for People Who (Think They) Hate Statistics, 4th ed. Sage, Thousand Oaks, CA.

Sheppard, S. A., A. Wiggins, and L. G. Terveen. 2014. Capturing quality: retaining provenance for curated volunteer monitoring data. Pages 1234–1245 in CSCW 2014—Proceedings of the 17th ACM Conference on Computer Supported Cooperative Work and Social Computing. Association for Computing Machinery.

Strasser, C., R. Cook, W. Michener, and A. Budden. 2011. Primer on data management: what you always wanted to know. DataONE, Albuquerque, NM.

Sutherland, W. J., A. S. Pullin, P. M. Dolman, and T. M. Knight. 2004. The need for evidence-based conservation. Trends in Ecology & Evolution 19:305–308.

University of Virginia Libraries. 2019. Steps in the data life cycle. University of Virginia Libraries: Research Data Services + Sciences. Obtained from https://data.library.virginia.edu/data-management/lifecycle/. Accessed October 27, 2019.

Wiggins, A., R. Bonney, E. Graham, S. Henderson, S. Kelling, R. Littauer, G. LeBuhn, K. Lotts, W. Michener, G. Newman, E. Russell, R. Stevenson, and J. Weltzin. 2013. Data management guide for public participation in scientific research. DataONE, Albuquerque, NM. www.dataone.org/sites/all/documents/DataONE-PPSR-DataManagementGuide.pdf.

Wiggins, A., R. D. Stevenson, G. Newman, and K. Crowston. 2011. Mechanisms for data quality and validation in citizen science. 2011 IEEE Seventh International Conference on eScience Workshops.

Bell, S., M. Marzano, J. Cent, H. Kobierska, D. Podjed, D. Vandzinskaite, H. Reinert, A. Armaitiene, M. Grodzińska-Jurczak, and R. Muršič. 2008. What counts? Volunteers and their organisations in the recording and monitoring of biodiversity. Biodiversity and Conservation 17:3443–3454.

Caldwell, W., and K. Oberhauser. 2013. Monarch Breeding Habitat Assessment Tool. www .monarchjointventure.org/images/uploads/documents/habitat_assessment_tool_final_ test.pdf.

Cooper, C. B., J. Shirk, and B. Zuckerberg. 2015. The invisible prevalence of citizen science in global research: migratory birds and climate change. PLoS ONE 9:e106508.

Davis, A. K. 2015. Opinion: conservation of monarch butterflies (Danaus plexippus) could be enhanced with analyses and publication of citizen science tagging data. Insect Conservation and Diversity 8:103–106.

Gewin, V. 2014. Science and politics: hello, governor. Nature 511:402–404.

Hoover, E. 2016. "We're not going to be guinea pigs": citizen science and environmental health in a Native American community. Journal of Science Communication 15:A05.

Jenkins, M. 2014. On the wing. Nature Conservancy Magazine, August-September, 50–59.

Lewandowski, E. J., and K. S. Oberhauser. 2017. Butterfly citizen scientists in the United States increase engagement in conservation. Biological Conservation 208:106–112.

Meyer, J. L., P. C. Frumhoff, S. P. Hamburg, and C. de la Rosa. 2010. Above the din but in the fray: environmental scientists as effective advocates. Frontiers in Ecology and the Environment 8:299–305.

Monarch Larva Monitoring Project. 2014. http://monarchlab.org/images/uploads/mlmp_ resources/Working_with_the_Media_MLMP.pdf.

Perrin, A., and M. Duggan. 2015. American's Internet access: 2000–2015. Pew Research Center, Washington, DC.

Ries, L., and K. S. Oberhauser. 2015. A citizen-army for science: quantifying the contributions of citizen scientists to our understanding of monarch butterfly biology. BioScience 65:419–430.

Riesch, H., and C. Potter. 2014. Citizen science as seen by scientists: methodological, epistemological and ethical dimensions. Public Understanding of Science 23:107–120.

Rotman, D., J. Preece, J. Hammock, K. Procita, D. Hansen, C. Parr, D. Lewis, and D. Jacobs. 2012. Pages 217–226 in Proceedings of the ACM 2012 Conference on Computer Supported Cooperative Work, Seattle, WA.

Roy, H. E., M. J. O. Pocock, C. D. Preston, D. B. Roy, J. Savage, J. C. Tweddle, and L. D. Robinson. 2012. Understanding citizen science and environmental monitoring. Final report on behalf of UK-EOF. NERC Centre for Ecology & Hydrology and Natural History Museum, UK. www.ukeof.org.uk/resources/citizen-science-resources/.

Steffy, G. 2015. Trends observed in fall migrant monarch butterflies (Lepidoptera: Nymphalidae) east of the Appalachian Mountains at an inland stopover in southern Pennsylvania over an eighteen year period. Annals of the Entomological Society of America 108:718–728.

## CHAPTER 12. PROGRAM EVALUATION

Ballard, H. L. 2008. What makes a scientist? Studying the impacts of harvest in the Pacific Northwest, USA. Pages 98–114 in L. Fortmann (editor), Participatory Research for Conservation and Rural Livelihoods: Doing Science Together. Blackwell, Oxford, UK.

Baranowski, T. 1985. Methodological issues in self-report of health behavior. Journal of School Health 55:179–182.

Biel, A. 2003. Environmental behaviour: changing habits in a social context. Pages 11–25 *in* A. Biel, B. Hannson, and M. Mårtensson (editors), Individual and Structural Determinants of Environmental Practice. Ashgate, London.

Bonney, R., C. B. Cooper, J. L. Dickinson, S. Kelling, T. Phillips, K. V. Rosenberg, and J. Shirk. 2009. Citizen science: a developing tool for expanding science knowledge and scientific literacy. BioScience 59:977–984.

Brossard, D., B. Lewenstein, and R. Bonney. 2005. Scientific knowledge and attitude change: the impact of a citizen science project. International Journal of Science Education 27:1099–1121.

Cooper, C. B., J. Dickinson, T. Phillips, and R. Bonney. 2007. Citizen science as a tool for conservation in residential ecosystems. Ecology and Society 12:11–22.

Couvet, D., F. Jiguet, R. Julliard, H. Levrel, and A. Teyssedre. 2008. Enhancing citizen contributions to biodiversity science and public policy. Interdisciplinary Science Reviews 33:95–103.

Crall, A. W., R. C. Jordan, K. Holfelder, G. J. Newman, J. Graham, and D. M. Waller. 2013. The impacts of an invasive species citizen science training program on participant attitudes, behavior, and science literacy. Public Understanding of Science 22:745–764.

Crall, A. W., G. J. Newman, C. S. Jarnevich, T. J. Stohlgren, D. M. Waller, and J. Graham. 2010. Improving and integrating data on invasive species collected by citizen scientists. Biological Invasions 12:3419–3428.

Danielsen, F., N. D. Burgess, and A. Balmford. 2005. Monitoring matters: examining the potential of locally-based approaches. Biodiversity and Conservation 14:2507–2542.

Delaney, D. G., C. D. Sperling, C. S. Adams, and B. Leung. 2008. Marine invasive species: validation of citizen science and implications for national monitoring networks. Biological Invasions 10:117–128.

Devictor, V., R. J. Whittaker, and C. Beltrame. 2010. Beyond scarcity: citizen science programmes as useful tools for conservation biogeography. Diversity and Distributions 16:354–362.

Dhondt, A. A., D. L. Tessaglia, and R. L. Slothower. 1998. Epidemic mycoplasmal conjunctivitis in House Finches from eastern North America. Journal of Wildlife Diseases 34:265–280.

Dickinson, J. L., B. Zuckerberg, and D. N. Bonter. 2010. Citizen science as an ecological research tool: challenges and benefits. Annual Review of Ecology, Evolution, and Systematics 41:149–172.

Doran, G. T. 1981. There's a S.M.A.R.T. way to write management's goals and objectives. Management Review 70:35–36.

Evans, C., E. Abrams, R. Reitsma, K. Roux, L. Salmonsen, and P. P. Marra. 2005. The Neighborhood Nestwatch Program: participant outcomes of a citizen-science ecological research project. Conservation Education 19:589–594.

Farmer, R. G., M. L. Leonard, J. E. M. Flemming, and S. C. Anderson. 2014. Observer aging and long-term avian survey data quality. Ecology and Evolution 4:2563–2576.

Fitzpatrick, M., E. Preisser, A. Ellison, and J. Elkinton. 2009. Observer bias and the detection of low density populations. Applied Ecology 19:1673–1679.

Jordan, R. C., H. L. Ballard, and T. B. Phillips. 2012. Key issues and new approaches for evaluating citizen-science learning outcomes. Frontiers in Ecology and the Environment 10:307–309.

Jordan, R. C., W. R. Brooks, D. V. Howe, and J. G. Ehrenfeld. 2012. Evaluating the performance of volunteer mapping invasive plants in public conservation lands. Environmental Management 49:425–434.

Jordan, R. C., W. R. Brooks, A. Sorensen, and J. Ehrenfeld. 2014. Understanding plant invasions: an example of working with citizen scientists to collect environmental data. AIMS Environmental Science 1:38–44.

Jordan, R.C., S.A. Gray, D.V. Howe, W.R. Brooks, and J.G. Ehrenfeld. 2011. Knowledge gain and behavioral change in citizen-science programs. Conservation Biology 25:1148–1154.

Kery, M., J.A. Royle, H. Schmid, M. Schaub, B. Volet, J.A. Hafliger, and N. Zbinden. 2010. Site-occupancy distribution modeling to correct population-trend estimates derived from opportunistic observations. Conservation Biology 24:1388–1397.

Lepczyk, C.A. 2005. Integrating published data and citizen science to describe bird diversity across a landscape. Journal of Applied Ecology 42:672–677.

Meinhold, J.L., and A.J. Malkus. 2005. Adolescent environmental behaviors: can knowledge, attitudes, and self-efficacy make a difference? Environment and Behavior 37:511–532.

Ottinger, G. 2009. Buckets of resistance: standards and the effectiveness of citizen science. Science Technology Human Values 35:244–270.

Pierce, B.A., and K.J. Gutzwiller. 2007. Interobserver variation in frog call surveys. Journal of Herpetology 41:424–429.

Root, T.L., J.T. Price, K.R. Hall, S.H. Schneider, C. Rosenzweig, and J.A. Pounds. 2003. Fingerprints of global warming on wild animals and plants. Nature 421:57–60.

Schmeller, D., P. Henry, R. Julliard, B. Gruber, J. Clobert, F. Dziock, S. Lengyel, P. Nowicki, E. Déri, E. Budrys, T. Kull, K. Tali, et al. 2009. Advantages of volunteer-based biodiversity monitoring in Europe. Conservation Biology 23:307–316.

Shirk, J.L., H.L. Ballard, C.C. Wilderman, T. Phillips, A. Wiggins, R. Jordan, E. McCallie, M. Minarchek, B.V. Lewenstein, M.E. Krasny, and R. Bonney. 2012. Public participation in scientific research: a framework for deliberate design. Ecology and Society 17(2):29.

Silvertown, J., L. Cook, R. Cameron, M. Dodd, K. McConway, J. Worthington, P. Skelton, C. Anton, O. Bossdorf, B. Baur, M. Schilthuizen, B. Fontaine, et al. 2011. Citizen science reveals unexpected continental-scale evolutionary change in a model organism. PLoS ONE 6:e18927.

Snall, T.O. Kindvall, J. Nilsson, and T. Part. 2011. Evaluating citizen-based presence data for bird monitoring. Biological Conservation 144:804–810.

Stern, P.C. 2000. Towards a coherent theory of environmentally significant behavior. Environment and Behavior 27:723–743.

Thornton, T., and J. Leahy. 2012. Trust in citizen science research: a case study of the Groundwater Education Through Water Evaluation and Testing program. Journal of the American Water Resources Association 48:1032–1040.

Trumbull, D.J., R. Bonney, D. Bascom, and A. Cabral. 2000. Thinking scientifically during participation in a citizen-science project. Science Education 84:265–275.

Wilderman, C.C., A. Barron, and L. Imgrund. 2004. Top down or bottom up? ALLARM's experience with two operational models for community science. Article 235 in Proceedings of the 4th National Monitoring Conference, May 17–20, 2004, Chattanooga, TN.

Worthington, J.P., J. Silvertown, L. Cook, R. Cameron, M. Dodd, R.M. Greenwood, K. McConway, and P. Skelton. 2011. Evolution MegaLab: a case study in citizen science methods. Methods in Ecology and Evolution 3:303–309.

## CHAPTER 13. HOW PARTICIPATION IN CITIZEN SCIENCE PROJECTS IMPACTS INDIVIDUALS

Adger, W.N. 2003. Social capital, collective action and adaptation to climate change. Economic Geography 79:387–404.

Bonney, R. 2007. Citizen science at the Cornell Laboratory of Ornithology. Pages 213–229 in R.E. Yager and J.H. Falk (editors), Exemplary Science in Informal Education Settings: Standards-Based Success Stories. NSTA Press, Arlington, VA.

Bonney, R., T. B. Phillips, H. L. Ballard, and J. W. Enck. 2016. Can citizen science enhance public understanding of science? Public Understanding of Science 25:2–16.

Brossard, D., B. Lewenstein, and R. Bonney. 2005. Scientific knowledge and attitude change: the impact of a citizen science project. International Journal of Science Education 27: 1099–1121.

Carr, A. J. L. 2004. Why do we all need community science? Society & Natural Resources 17:841–849.

Conrad, C. T., and T. Daoust. 2008. Community-based monitoring frameworks: increasing the effectiveness of environmental stewardship. Environmental Management 41:358–366.

Crabbe, M. J. C. 2012. From citizen science to policy development on the coral reefs of Jamaica. International Journal of Zoology 2012: article 102350.

Evans, C., E. Abrams, R. Reitsma, K. Roux, L. Salmonsen, and P. P. Marra. 2005. The Neighborhood Nestwatch Program: participant outcomes of a citizen-science ecological research project. Conservation Biology 19:589–594.

Fernandez-Gimenez, M. E., H. L. Ballard, and V. E. Sturtevant. 2008. Adaptive management and social learning in collaborative and community-based monitoring: a study of five community-based forestry organizations in the western USA. Ecology and Society 13:4.

Gardiner, M. M., L. L. Allee, P. M. J. Brown, J. E. Losey, H. E. Roy, and R. R. Smyth. 2012. Lessons from lady beetles: accuracy of monitoring data from US and UK citizen-science programs. Frontiers in Ecology and the Environment 10:471–476.

Genet, K. S., C. A. Lepczyk, R. Christoffel, L. G. Sargent, and T. M. Burton. 2008. Using volunteer monitoring programs for anuran conservation along a rural-urban gradient in southern Michigan, USA. Pages 565–574 in R. E. Jung and J. C. Mitchell (editors), Urban Herpetology: Ecology, Conservation and Management of Amphibians and Reptiles in Urban and Suburban Environments. Society for the Study of Amphibians and Reptiles, Salt Lake City, UT.

Homayoun, T. Z., and R. B. Blair. 2008. Citizen-science monitoring of landbirds in the Mississippi River Twin Cities Important Bird Area. Pages 607–616 in Proceedings of the Fourth International Partners in Flight Conference: Tundra to Tropics.

Jolly, E. C. 2002. Confronting demographic denial: retaining relevance in the new millennium. Journal of Museum Education 27:2–4.

Jordan, R. C., S. A. Gray, D. V. Howe, W. R. Brooks, and J. G. Ehrenfeld. 2011. Knowledge gain and behavioral change in citizen-science programs. Conservation Biology 25:1148–1154.

Knutson, M. G., J. R. Sauer, D. A. Olsen, M. J. Mossman, L. M. Hemesath, and M. J. Lannoo. 1999. Effects of landscape composition and wetland fragmentation on frog and toad abundance and species richness in Iowa and Wisconsin, USA. Conservation Biology 13:1437–1446.

Kountoupes, D. L., and K. S. Oberhauser. 2012. Citizen science and youth audiences: educational outcomes of the Monarch Larva Monitoring Project. Journal of Community Engagement and Scholarship 1:10–20.

Leslie, L. L., C. E. Velez, and S. A. Bonar. 2004. Utilizing volunteers on fisheries projects: benefits, challenges, and management techniques. Fisheries 29:10–14.

Nerbonne, J. F., and K. C. Nelson. 2004. Volunteer macroinvertebrate monitoring in the United States: resource mobilization and comparative state structures. Society & Natural Resources 17:817–839.

Overdest, C., C. H. Orr, and K. Stepenuck. 2004. Volunteer stream monitoring and local participation in natural resource issues. Human Ecology Review 11:177–185.

Pandya, R. E. 2012. A framework for engaging diverse communities in citizen science in the US. Frontiers in Ecology and the Environment 10:314–317.

Price, C. A., and H.-S. Lee. 2013. Changes in participants' scientific attitudes and epistemological beliefs during an astronomical citizen science project. Journal of Research in Science Teaching 50:773–801.

Rohs, F. R., J. H. Stribling, and R. R. Westerfield. 2002. What personally attracts volunteers to the Master Gardener program? Journal of Extension 40. www.joe.org/joe/2002august /rb5.php.

Shirose, L. J., C. A. Bishop, D. M. Green, C. J. MacDonald, R. J. Brooks, and N. J. Helferty. 1997. Validation tests of an amphibian call count survey technique in Ontario, Canada. Herpetologica 53:312–320.

Smith, W. H., S. L. Slemp, C. D. Stanley, M. N. Blackburn, and J. Wayland. 2015. Combining citizen science with traditional biotic surveys to enhance knowledge regarding the natural history of secretive species: notes on the geographic distribution and status of the Green Salamander (*Aneides aeneus*) in the Cumberland Mountains of Virginia, USA. IRCF Reptiles and Amphibians 22:135–144.

Thompson, S., and R. Bonney. 2007. Evaluating the impacts of participation in an online citizen science project: a mixed-methods approach. Pages 187–199 *in* J. Trant and D. Bearman (editors), Museums and the Web 2007: Proceedings. Archives & Museum Informatics, Toronto.

Trumbull, D. J., R. Bonney, D. Bascom, and A. Cabral. 2000. Thinking scientifically during participation in a citizen-science project. Science Education 84:265–275.

Trumbull, D. J., R. Bonney, and N. Grudens-Schuck. 2005. Developing materials to promote inquiry: lessons learned. Science Education 89:879–900.

Wilderman, C. C., A. Barron, and L. Imgrund. 2004. Top down or bottom up? ALLARM's experience with two operational models for community science. Article 235 *in* Proceedings of the 4th National Monitoring Conference, May 17–20, 2004, Chattanooga, TN.

CHAPTER 14. FROM TINY ACORNS GROW MIGHTY OAKS

Ault, T. R., G. M. Henebry, K. M. deBuers, M. D. Schwartz, J. L. Betancourt, and D. Moore. 2013. The false spring of 2012, earliest in North America record. Eos 94:181–183.

Betancourt, J. L., M. D. Schwartz, D. D. Breshears, C. A. Brewer, G. Frazer, J. E. Gross, S. J. Mazer, B. C. Reed, and B. E. Wilson. 2007. Evolving plans for the USA National Phenology Network. Eos 88:211–211.

Betancourt, J. L., M. D. Schwartz, D. D. Breshears, D. R. Cayan, M. D. Dettinger, D. W. Inouye, E. Post, and B. C. Reed. 2005. Implementing a U.S. National Phenology Network. Eos 86:539–539.

Boudreau, S. A., and N. D. Yan. 2004. Auditing the accuracy of a volunteer-based surveillance program for an aquatic invader *Bythotrephes*. Environmental Monitoring and Assessment 91:17–26.

Cash, D., W. C. Clark, F. Alcock, N. M. Dickson, N. Ecklye, and J. Jäger. 2002. Salience, credibility, legitimacy and boundaries: linking research, assessment and decision making. Working Paper Series rwp02–046, Harvard University, John F. Kennedy School of Government.

Crimmins, T. M., J. F. Weltzin, A. H. Rosemartin, E. M. Surina, L. Marsh, and E. G. Denny. 2014. Targeted campaign increases activity among participants in Nature's Notebook, a citizen science project. Natural Sciences Education 43:64–72.

Darling-Hammond, L., B. Barron, P. D. Pearson, A. H. Schoenfeld, E. K. Stage, T. D. Zimmerman, G. N. Cervetti, and J. L. Tilson. 2008. Powerful Learning: What We Know about Teaching for Understanding. Wiley, Hoboken, NJ.

Denny, E. G., K. L. Gerst, A. J. Miller-Rushing, G. L. Tierney, T. M. Crimmins, C. A. F. Enquist, P. Guertin, A. H. Rosemartin, M. D. Schwartz, K. A. Thomas, and J. F. Weltzin. 2014. Standardized phenology monitoring methods to track plants and animal activity for science and resource management applications. International Journal of Biometeorology 58:591–601.

Dickinson, J. L., B. Zuckerberg, and D. N. Bonter. 2010. Citizen science as an ecological research tool: challenges and benefits. Annual Review of Ecology and Systematics 41:49–72.

Elmendorf, S. C., K. D. Jones, B. I. Cook, J. M. Diez, C. A. F. Enquist, R. A. Hufft, M. O. Jones, S. J. Mazer, A. J. Miller-Rushing, D. J. P. Moore, M. D. Schwartz, and J. F. Weltzin. 2016. The plant phenology monitoring design for the National Ecological Observatory Network. Ecosphere 7:e01303.

Elmore, A. J., C. D. Stylinski, and K. Pradhan. 2016. Synergistic use of citizen science and remote sensing for continental-scale measurements of forest tree phenology. RemoteSensing 8:502.

Enquist, C. A. F., A. Rosemartin, and M. D. Schwartz. 2012. Identifying and prioritizing phenological data products and tools. Eos 93:356.

Fuccillo, K. K., T. M. Crimmins, C. DeRivera, and T. S. Elder. 2015. Assessing accuracy in volunteer-based plant phenology monitoring. International Journal of Biometeorology 59:917–926.

Glynn, P. D., and T. W. Owen, editors. 2015. Review of the USA National Phenology Network. U.S. Geological Survey Circular 1411. http://dx.doi.org/10.3133/cir1411.

Guertin, P., L. Barnett, E. G. Denny, and S. N. Schaffer. 2015. USA National Phenology Network Botany Primer. USA-NPN Education and Engagement Series 2015–001. Version 1.1. www .usanpn.org/files/shared/files/USA-NPN_Botany-Primer.pdf.

Guston, D. H. 2001. Boundary organizations in environmental policy and science: an introduction. Science, Technology, & Human Values 26:399–408.

Hampton, S. E., C. A. Strasser, J. J. Tewksbury, W. K. Gram, A. E. Budden, A. L. Batcheller, C. S. Duke, and J. H. Porter. 2013. Big data and the future of ecology. Frontiers in Ecology and the Environment 11:156–162.

Kellogg Foundation. 2004. Logic model development guide. Kellogg Foundation, Battle Creek, MI.

McKinley, D. C., A. J. Miller-Rushing, H. L. Ballard, R. E. Bonney, H. Brown, S. Cook-Patton, D. M. Evans, R. A. French, J. K. Parrish, T. B. Phillips, S. F. Ryan, L. A. Shanley, et al. 2017. Citizen science can improve conservation science, natural resource management, and environmental protection. Biological Conservation 208:15–28.

Michener, W. K., and M. B. Jones. 2012. Ecoinformatics: supporting ecology as a data-intensive science. Trends in Ecology & Evolution 27:85–93.

Miller, K. L. 2013. Content Marketing for Nonprofits. Jossey-Bass, San Francisco, CA.

Nolan, V. P., and J. F. Weltzin. 2011. Phenology for science, resource management, decision making, and education. Eos 92:15.

Parmesan, C. 2007. Influences of species, latitudes and methodologies on estimates of phenological response to global warming. Global Change Biology 13:1860–1872.

Peters, D. P. C., K. M. Havstad, J. Cushing, C. Tweedie, O. Fuentes, and N. Villanueva-Rosales. 2014. Harnessing the power of big data: infusing the scientific method with machine learning to transform ecology. Ecosphere 5:67.

Rosemartin, A. H., E. G. Denny, K. L. Gerst, R. L. Marsh, T. M. Crimmins, and J. F. Weltzin. 2018a. USA National Phenology Network observational data documentation. U.S. Geological Survey Open-File Report 2018–1060. https://doi.org/10.3133/ofr20181060.

Rosemartin, A., E. G. Denny, J. F. Weltzin, R. L. Marsh, B. E. Wilson, H. Mehdipoor, R. Zurita-Milla, and M. D. Schwartz. 2015. Lilac and honeysuckle phenology data 1956–2014. Scientific Data 2:150038.

Rosemartin, A., M. L. Langseth, T. M. Crimmins, and J. F.Weltzin. 2018b. Development and release of phenological data products—a case study in compliance with federal open data policy. U.S. Geological Survey Open-File Report 2018–1007. https://doi.org/10.3133/ofr20181007.

Schwartz, M. D., J. L. Betancourt, and J. F. Weltzin. 2012. From Caprio's lilacs to the USA National Phenology Network. Frontiers in Ecology and the Environment 10:324–327.

Soranno, P. A., and D. S. Schimel. 2014. Macrosystems ecology: big data, big ecology. Frontiers in Ecology and the Environment 12:3.

Taylor, S. D., J. M. Meiners, K. Riemer, M. C. Orr, and E. P. White. 2019. Comparison of large-scale citizen science data and long-term study data for phenology modeling. Ecology 100:e02568.

Theobald, E. J., A. K. Ettinger, H. K. Burgess, L. B. DeBey, N. R. Schmidt, H. E. Froehlich, C. Wagner, J. HilleRisLambers, J. Tewksbury, M. A. Harsch, and J. K. Parrish. 2015. Global change and local solutions: tapping the unrealized potential of citizen science for biodiversity research. Biological Conservation 181:236–244.

Tierney, G., B. Mitchell, A. Miller-Rushing, J. Katz, E. G. Denny, C. Brauer, T. Donovan, A. D. Richardson, M. Toomey, A. Kozlowski, J. Weltzin, K. Gerst, et al. 2013. Phenology monitoring protocol: Northeast Temperate Network. Natural Resource Report NPS/NETN/NRR—2013/681. National Park Service, Fort Collins, CO.

USA-NPN National Coordinating Office. 2014. USA National Phenology Network five-year strategic plan (F14-FY18). USA-NPN Programmatic Series 2014–001. www.usanpn.org.

USA-NPN National Coordinating Office. 2016. Data quality assurance & quality control for Nature's Notebook. Technical Information Sheet. Version 2.3.

Wenger, E. 1998. Communities of Practice: Learning, Meaning, and Identity. Cambridge University Press, Cambridge, UK.

Wiggins, A., R. Bonney, E. Graham, S. Henderson, S. Kelling, R. Littauer, G. LeBuhn, K. Lotts, W. Michener, G. Newman, E. Russell, R. Stevenson, and J. Weltzin. 2013. Data management guide for public participation in scientific research. DataONE, Albuquerque, NM. www.dataone.org/sites/all/documents/DataONE-PPSR-DataManagementGuide.pdf.

Yue, X., N. Unger, T. Keenan, X. Zhang, and C. S. Vogel. 2015. Probing the past 30-year phenology trend of U.S. deciduous forests. Biogeosciences 12:4693–4709.

## CHAPTER 15. CITIZEN SCIENCE AT THE URBAN ECOLOGY CENTER

Barlow, D. 2006. Urban Ecology Center trail study. Unpublished intern report for the Urban Ecology Center, Milwaukee, WI.

Beyer, K., E. Heller, J. Bizub, A. Kistner, A. Szabo, E. Shawgo, and C. Zetts. 2015. More than a pretty place: assessing the impact of environmental education on children's knowledge and attitudes about outdoor play in nature. International Journal of Environmental Research and Public Health 12:2054–2070.

Chawla, L. 1998. Significant life experiences revisited: a review of research on sources of environmental sensitivity. Environmental Education Research 4:369–382.

Runyard, A. 2005. Implementation of a geographic information system at the Urban Ecology Center. Riverside Park, Milwaukee, Wisconsin. Unpublished intern report for the Urban Ecology Center, Milwaukee, WI.

## CHAPTER 16. DRIVEN TO DISCOVER

Akerson, V., and L. A. Donnelly. 2010. Teaching nature of science to K–12 students: what understanding can they attain? International Journal of Science Education 32:97–124.

Archer, L., J. DeWitt, J. Osborne, J. Dillon, B. Willis, and B. Wong. 2010. "Doing" science versus "being" a scientist: examining 10- and 11-year-old schoolchildren's constructions of science through the lens of identity. Science Education 94:617–639.

Berland, L. K., and K. L. McNeill. 2010. A learning progression for scientific argumentation: understanding student work and designing supportive instructional contexts. Science Education 94:765–793.

Bonney, R., H. Ballard, R. Jordan, E. McCallie, T. Phillips, J. Shirk, and C. C. Wilderman. 2009. Public participation in scientific research: defining the field and assessing its potential for informal science education. A CAISE Inquiry Group report. CAISE, Washington, DC.

Bouillon, L. M., and L. M. Gomez. 2001. Connecting school and community with science learning: real world problems and school-community partnerships as contextual scaffolds. Journal of Research in Science Teaching 38:878–898.

Brossard, D., B. Lewenstein, and R. Bonney. 2005. Scientific knowledge and attitude change: the impact of a citizen science project. International Journal of Science Education 27: 1099–1121.

Chawla, L. 1990. Ecstatic places. Children's Environments Quarterly 7:18–23.

Ferry, B. 1995. Enhancing environmental experiences through effective partnership teacher educators, field study centers and schools. Journal of Experiential Education 18:133–137.

Fogleman, T., and M. C. Curran. 2008. How accurate are student-collected data? Science Teacher 75:30–35.

Global Learning and Observations to Benefit the Environment. 2014. The GLOBE program. www .globe.gov/.

Gootman, J. A., and J. Eccles, editors. 2002. Community Programs to Promote Youth Development. National Academies Press, Washington, DC.

Hulleman, C. S., and J. M. Harackiewicz. 2009. Promoting interest and performance in high school science classes. Science 326:1410.

Juhl, L., K. Yearsley, and A. J. Silva. 1997. Interdisciplinary project-based learning through an environmental water quality study. Journal of Chemical Education 74:1431–1433.

Kirch, S. 2010. Identifying and resolving uncertainty as a mediated action in science: a comparative analysis of the cultural tools used by scientists and elementary science students at work. Science Education 94:308–335.

Louv, R. 2005. Last Child in the Woods: Saving Our Children from Nature-Deficit Disorder. Algonquin Books, Chapel Hill, NC.

Metz, K. 2006. The knowledge-building enterprises in science and elementary school science classrooms. Pages 105–130 in L. B. Flick and N. G. Lederman (editors), Scientific Inquiry and Nature of Science. Kluwer Academic, Dordrecht, The Netherlands.

Meyer, N. J., S. Scott, A. L. Strauss, P. L. Nippolt, K. S. Oberhauser, and R. B. Blair. 2014. Citizen science as a REAL environment for authentic scientific inquiry. Journal of Extension 52:4.

Mueller, M. P., D. Tippins, and L. A. Bryan. 2012. The future of citizen science. Democracy & Education 20(1): article 2.

National Research Council. 2000. Inquiry and the National Science Education Standards: A Guide for Teaching and Learning. National Academies Press, Washington, DC.

Schaus, J. M., R. Bonney, A. J. Rosenberg, and C. B. Phillips. 2007. BirdSleuth: investigating evidence. Cornell Lab of Ornithology, Ithaca, NY.

Shirk, J. L., H. L. Ballard, C. C. Wilderman, T. Phillips, A. Wiggins, R. Jordan, E. McCallie, M. Minarchek, B. V. Lewenstein, M. E. Krasny, and R. Bonney. 2012. Public participation in scientific research: a framework for deliberate design. Ecology and Society 17(2):29.

Swail, W. S., and E. Kampits. 2004. Work-based learning & higher education: a research perspective. American Higher Education Report Series. Educational Policy Institute, Washington, DC.

Tan, E., and A. C. Barton. 2010. Transforming science learning and student participation in sixth grade science: a case study of a low-income, urban, racial minority classroom. Equity and Excellence in Education 43:38–55.

Trautmann, N. M., J. L. Shirk, J. Fee, and M. E. Krasny. 2012. Who poses the question? Using citizen science to help K–12 teachers meet the mandate for inquiry. Pages 179–190 *in* J. L. Dickinson and R. Bonney (editors), Citizen Science: Public Participation in Environmental Research. Cornell University Press, Ithaca, NY.

Trumbull, D. J., R. Bonney, D. Bascom, and A. Cabral. 2000. Thinking scientifically during participation in a citizen-science project. Science Education 84:265–275.

## CHAPTER 17. CHALLENGES OF FOREST CITIZEN INVOLVEMENT IN BIODIVERSITY MONITORING IN PROTECTED AREAS OF BRAZILIAN AMAZONIA

AMMAI-AC. 2007. Plano de gestão territorial e ambiental da Terra Indígena Kaxinawá/Asheninka do Rio Breu. Comissão Pró-Índio do Acre, Rio Branco, Brazil.

Apiwtxa. 2007. Plano de Gestão Ashaninka. Apiwtxa/AMAAI-AC/CPI-AC, Rio Branco, Brazil.

Benchimol, M., E. M. Muhlen, and E. M. Venticinque. 2017. Lessons from a community-based program to monitor forest vertebrates in the Brazilian Amazon. Environmental Management 60:476–483.

Constantino, P. A. L. 2015. Dynamics of hunting territories and prey distribution in Amazonian indigenous lands. Applied Geography 56:222–231.

Constantino, P. A. L. 2016. Deforestation and hunting effects on wildlife across in Amazonian indigenous lands. Ecology and Society 21(2):3.

Constantino, P. A. L. 2019. Subsistence hunting with mixed-breed dogs reduces hunting pressure on sensitive Amazonian game species in protected areas. Environmental Conservation 46:92–98.

Constantino, P. A. L., M. Benchimol, and A. P. Antunes. 2018. Designing indigenous lands in Amazonia: securing indigenous rights and wildlife conservation through hunting management. Land Use Policy 77:652–660.

Constantino, P. A. L., H. S. A. Carlos, E. E. Ramalho, L. Rostant, C. E. Marinelli, D. Teles, and S. F. Fonseca Jr. 2012a. Empowering local people through community-based resource monitoring: a comparison between Brazil and Namibia. Ecology and Society 17(4):22.

Constantino, P. A. L., L. B. Fortini, F. R. S. Kaxinawa, A. M. Kaxinawa, E. S. Kaxinawa, A. P. Kaxinawa, and L. S. Kaxinawa. 2008. Indigenous collaborative research for wildlife management in Amazonia: the case of the Kaxinawá, Acre, Brazil. Biological Conservation 141:2718–2729.

Constantino, P. A. L., R. A. Tavares, J. L. Kaxinawa, F. M. Macário, E. S. Kaxinawa, and A. S. Kaxinawa. 2012b. Mapeamento e monitoramento participativo da caça na Terra Indígena Kaxinawá da Praia do Carapanã, Acre, Amazônia Brasileira. Pages 141–153 *in* A. Paese, A. Uezu, M. L. Lorini, and A. Cunha (editors), Sistema de informações geográficas e a conservação da biodiversidade. Oficina do Texto, São Paulo, Brazil.

Danielsen, F., N. D. Burgess, A. Balmford, P. F. Donald, M. Funder, J. P. G. Jones, P. Alviola, D. S. Balete, T. Blomley, J. Brashares, B. Child, M. Enghoff, et al. 2009. Local participation in natural resource monitoring: a characterization of approaches. Conservation Biology 23:31–42.

Evans, K., and M. R. Guariguata. 2008. Participatory monitoring in tropical forest management: a review of tools, concepts and lessons learned. CIFOR, Bogor, Indonesia.

Gomes, A. S. R. 2017. Automonitoramento Paiter Surui sobre o uso de mamíferos de médio e grande porte na terra indígena Sete de Setembro, Cacoal, Rondônia, Brasil. ECAM, Porto Velho, Brazil.

Hecht, S. 2011. The new Amazon geographies: insurgent citizenship, "Amazon nation" and the politics of environmentalism. Journal of Cultural Ecology 28:203–223.

IBAMA (Instituo Brasileiro de Meio Ambiente e dos Recursos Naturais). 2006. I Seminário de monitoramento da biodiversidade em unidades de proteção integral da Amazônia. FUNBIO and ARPA, Brasília DF, Brazil.

IBGE (Instituto Brasileiro de Geografia e Estatística). 2013. Síntese de indicadores sociais: uma análise das condições de vida da população brasileira. ftp://ftp.ibge.gov.br/Indicadores_Sociais/Sintese_de_Indicadores_Sociais_2013/pdf/educacao_pdf.pdf.

ICMBio (Instituto Chico Mendes de Conservação da Biodiversidade). 2014. ICMBio em foco. Edição 279. Brasília DF, Brazil.

ISA (Instituto Socioambiental). 2019. Localização e extensão das TIs. https://pib.socioambiental .org/pt/Localiza%C3%A7%C3%A3o_e_extens%C3%A3o_das_TIs.

Matos, K. G., and N. L. Monte. 2006. O estado da arte da formação de professores indígenas no Brasil in Grupioni, LDB (Org.) Formação de professores indígenas: repensando trajetórias. Ministério da Educação. Brasília DF, Brazil.

MMA (Ministério do Meio Ambiente). 2012. Estratégia nacional para conservação e uso sustentável da biodiversidade brasileira: ampliação e consolidação do sistema nacional de unidades de conservação da natureza 2012–2020. MMA, Brasília DF, Brazil.

Monte, N. L. 2000. The others, who we are? Training indigenous teachers and intercultural identities. Caderno de Pesquisa 111:7–29.

Pereira, R. C., F. O. Roque, P. A. L. Constantino, J. Sabino, and M. Uehara-Prado. 2013. Monitoramento in situ da biodiversidade: Proposta para um sistema brasileiro de monitoramento da biodiversidade. ICMBio, Brasília DF, Brazil.

Santos, R. S. S, A. B. Pereira, T. Pereira, J. Pereira, F. Prado, and P. A. L. Constantino. 2014. Estrutura pedagógica do ciclo de capacitação em monitoramento da biodiversidade. ICMBio, Brasília DF, Brazil.

Schmink, M. 2011. Forest citizens: changing life conditions and social identities in the land of rubber tappers. Latin American Research Review 46:141–158.

CHAPTER 18. DOCUMENTING THE CHANGING LOUISIANA WETLANDS
THROUGH COMMUNITY-DRIVEN CITIZEN SCIENCE

Barbier, E., E. Koch, B. Silliman, S. Hacker, E. Wolanski, J. Primavera, E. Granek, S. Polasky, S. Aswani, L. Cramer, D. Stoms, C. Kennedy, et al. 2008. Coastal ecosystem-based management with nonlinear ecological functions and values. Science 319: 321–323.

Batker, D., I.de la Torre, R. Costanza, P. Swedeen, J. Day, R. Boumans, and K. Bagstad. 2010. Gaining ground: wetlands, hurricanes, and the economy: the value of restoring the Mississippi River Delta. Earth Economics, Tacoma, WA.

Batker, D., I. de la Torre, R. Costanza, P. Swedeen, J. Day, R. Boumans, and K. Bagstad. 2014. The threats to the value of ecosystem goods and services of the Mississippi Delta. Pages 155–173 in J. Day, G. Kemp, A. Freeman, and D. Muth (editors), Perspectives on the Restoration of the Mississippi Delta. Springer, Dordrecht, The Netherlands.

Coastal Protection and Restoration Authority. 2012. Louisiana's Comprehensive Master Plan for a Sustainable Coast Appendix A2—Project Fact Sheets. http://coastal.la.gov/our-plan/2012-coastal-masterplan/cmp-appendices/.

Coastal Protection and Restoration Authority. 2014. Integrated Ecosystem Restoration & Hurricane Protection in Coastal Louisiana: Fiscal Year 2015 Annual Plan. http://coastal.la.gov/wp-content/uploads/2014/01/AP_web1.pdf.

Condrey, R. E., P. Hoffman, and D. E. Evers. 2014. The last naturally active delta complexes of the Mississippi River (LNDM): discovery and implications. Pages 33–50 in J. Day, G. Kemp, A. Freeman, and D. Muth (editors), Perspectives on the Restoration of the Mississippi Delta. Springer, Dordrecht, The Netherlands.

Couvillion, B. R., J. A. Barras, G. D. Steyer, W. Sleavin, M. Fischer, H. Beck, N. Trahan, B. Griffin, and D. Heckman. 2011. Land area change in coastal Louisiana from 1932 to 2010: U.S. Geological Survey Scientific Investigations Map 3164, scale 1:265,000.

Dalbom, C., S. Hemmerling, and J. Lewis. 2014. Community resettlement prospects in southeast Louisiana: a multidisciplinary exploration of legal, cultural, and demographic aspects of moving individuals and communities. White paper, Tulane Institute on Water Resources Law & Policy, New Orleans, LA.

Dosemagen, S., A. Ameen, and A. Kolker. 2011. Barataria Site BH1. Wilkinson Bayou, Louisiana. http://publiclab.org/map/wilkinson-bayou-louisiana-october-4/2011-10-4.

Environmental Protection Agency OIG. 2011. EPA must improve oversight of state enforcement. Report no. 12-P-0113.

Long, S. 2014a. Barataria Site BL1 Barataria Bay, Louisiana. Public Lab map archive. http://publiclab.org/map/barataria-site-bl1-barataria-bay-louisiana/04-19-2014.

Long, S. 2014b. Pelican Point. Barataria Bay, Louisiana. Public Lab map archive. http://publiclab.org/map/pelican-point-barataria-bay-louisiana/04-19-2014.

National Academy of Sciences. 2009. The New Orleans Hurricane Protection System: Assessing Pre-Katrina Vulnerability and Improving Mitigation and Preparedness. National Academies Press, Washington, DC.

Ward, W. 2013. Cartography primer—Pamet Marsh. Public Lab Research Note. http://publiclab.org/notes/wward1400/11-07-2013/cartography-primer-pamet-marsh.

Wirth, G. 2012. Yellow Bar Island—dredge reuse and FAA permitting. Public Lab Research Note. http://publiclab.org/notes/gwirth/7-23-2012/yellow-bar-island-dredge-reuse.

## CHAPTER 19. REEF CHECK CALIFORNIA

Bond, N. A., M. F. Cronin, H. Freeland, and N. Mantua. 2015. Causes and impacts of the 2014 warm anomaly in the NE Pacific. Geophysical Research Letters 42:3414–3420.

Byrnes, J., J. J. Stachowicz, K. M. Hultgren, A. Randall Hughes, S. V. Olyarnik, and C. S. Thornber. 2006. Predator diversity strengthens trophic cascades in kelp forests by modifying herbivore behaviour. Ecology Letters 9:61–71.

California Department of Fish and Game. 2008. Master plan for Marine Protected Areas. California Department of Fish and Game, Sacramento.

California Fish and Game Code. (Marine Life Protection Act) Sections 2850–2863.

Carr, M., and D. Reed. 2016. Shallow rocky reefs and kelp forests. Pages 311–336 in H. Mooney and E. Zavaleta (editors), Ecosystems of California. University of California Press, Berkeley.

Clarke, K. R., and R. N. Gorley. 2006. PRIMER v6: user manual/tutorial. PRIMER-E, Plymouth, UK.

Dayton, P. K., M. J. Tegner, P. E. Parnell, and P. B. Edwards. 1992. Temporal and spatial patterns of disturbance and recovery in a kelp forest community. Ecological Monographs 62:421–445.

Foster, M. 1990. Organization of macroalgal assemblages in the northeast Pacific: the assumption of homogeneity and the illusion of generality. Hydrobiologia 192:21–33.

Foster, M. S., and D. R. Schiel. 1985. The ecology of giant kelp forests in California: a community profile. U.S. Fish and Wildlife Service Biological Report 85 (7.2).

Freiwald, J., and M. Wehrenberg. 2013. Reef Check California: north Central Coast baseline surveys of shallow rocky reef ecosystems. Reef Check Foundation, Pacific Palisades, CA.

Freiwald, J., and C. Wisniewski. 2015. Reef Check California: citizen scientist monitoring of rocky reefs and kelp forests: creating a baseline for California's South Coast. Final report, South Coast MPA Baseline Monitoring 2011–2014. Reef Check Foundation, Pacific Palisades, CA.

Freiwald, J., C. Wisniewski, M. Wehrenberg, C. S. Shuman, and C. Dawson. 2015. Reef Check California Instruction Manual: A Guide to Rocky Reef Monitoring, 8th ed. Reef Check Foundation, Pacific Palisades, CA.

Gillett, D., D. Pondella II, J. Freiwald, K. Schiff, J. Caselle, C. Shuman, and S. Weisberg. 2012. Comparing volunteer and professionally collected monitoring data from the rocky subtidal reefs of Southern California, USA. Environmental Monitoring and Assessment 184:3239–3257.

Graham, M. H. 2004. Effects of local deforestation on the diversity and structure of Southern California giant kelp forest food webs. Ecosystems 7:341–357.

Hewson, I., J. B. Button, B. M. Gudenkauf, B. Miner, A. L. Newton, J. K. Gaydos, J. Wynne, C. L. Groves, G. Hendler, M. Murray, S. Fradkin, M. Breitbart, et al. 2014. Densovirus associated with sea-star wasting disease and mass mortality. Proceedings of the National Academy of Sciences USA 111:17278–17283.

Kirlin, J., M. Caldwell, M. Gleason, M. Weber, J. Ugoretz, E. Fox, and M. Miller-Henson. 2013. California's Marine Life Protection Act Initiative: supporting implementation of legislation establishing a statewide network of marine protected areas. Ocean & Coastal Management 74:3–13.

Ling, S. D., C. R. Johnson, S. D. Frusher, and K. R. Ridgway. 2009. Overfishing reduces resilience of kelp beds to climate-driven catastrophic phase shift. Proceedings of the National Academy of Sciences USA 106:22341–22345.

Marks, L., P. Salinas-Ruiz, D. Reed, S. Holbrook, C. Culver, J. Engle, D. Kushner, J. Caselle, J. Freiwald, J. Williams, J. Smith, L. Aguilar-Rosas, and N. Kaplanis. 2015. Range expansion of a non-native, invasive macroalga *Sargassum horneri* (Turner) C. Agardh, 1820 in the eastern Pacific. BioInvasions Records 4:243–248.

Moitoza, D. J., and D. W. Phillips. 1979. Prey defense, predator preference, and nonrandom diet: the interactions between *Pycnopodia helianthoides* and two species of sea urchins. Marine Biology 53:299–304.

OST and CDFW. 2013. State of the California Central Coast: results from baseline monitoring of Marine Protected Areas 2007–2012. California Ocean Science Trust and California Department of Fish and Wildlife, Sacramento.

Pearse, J. S. 2006. Ecological role of purple sea urchins. Science 314:940–941.

Pearse, J. S., and A. H. Hines. 1987. Long-term population dynamics of sea urchins in a central California kelp forest: rare recruitment and rapid decline. Marine Ecology Progress Series 39:275–283.

Sayce, K., C. Shuman, D. Connor, A. Reisewitz, E. Pope, M. Miller-Henson, E. Poncelet, D. Monié, and B. Owens. 2013. Beyond traditional stakeholder engagement: public participation roles in California's statewide marine protected area planning process. Ocean & Coastal Management 74:57–66.

Schiel, D. R., and M. S. Foster. 2006. The population biology of large brown seaweeds: ecological consequences of multiphase life histories in dynamic coastal environments. Annual Review of Ecology, Evolution, and Systematics 37:343–372.

Springer, Y., C. Hays, M. Carr, and M. Mackey. 2006. Ecology and management of the bull kelp, *Nereocystis luetkeana:* a synthesis with recommendations for future research. Lenfest Ocean Program, Washington, DC.

Watanabe, J. M., and C. Harrold. 1991. Destructive grazing by sea urchins *Strongylocentrotus* spp. in a Central California kelp forest: potential roles of recruitment, depth, and predation. Marine Ecology Progress Series 71:125–141.

Weible, C. M. 2008. Caught in a maelstrom: implementing California Marine Protected Areas. Coastal Management 36:350–373.

# INDEX

Note: *Italic* page numbers indicate illustrations.

terms, 219; definitions of, 7, 14, 17; diversity of research in, 7; funding for, 2, 37–38; history of, 17–23; inappropriate uses of, 1, 9–11; need for, xiii; as scientific vs. educational tool, 224; skepticism about, xiv, 133, 163, 164, 219–220; strengths of, 8–9, 12–14

Citizen Science Association, 61, 71, 90, 182

Citizen Science Central, 90

citizen scientists: behavioral changes in, 177, 180, 181–182, 187; demographics of, 85–86, 185–186, 188, 264; ecological literacy of, xiii–xiv, 1–2, 22; gaps in knowledge about, 188; identifying potential, 38–40; motivations of (*See* motivations); personal benefits to, xv, 87, 185–188; recruitment of (*See* recruitment); as research subjects, 59–60; retention of (*See* retention); terms used for, 3, 162; underrepresented groups as, 39–40, 185–186, 188

Citizens Researching Invertebrate Kritters Together (CRIKT), 219

CitSci.org: communication tools of, 102; in data collection, 132; in data standardization, 121–122, 150; in project evaluation, 182; summary of tools of, *144;* Urban Ecology Center and, 214

Clark, William, 20

Clean Water Act, 58

Cleveland, North Chagrin Nature Center in, *232, 233*

clickwrap user agreements, 64–65

climate change, historical collections and records in study of, 19–20

ClimateWatch, *144*

cloud resources, *145, 149*

Coastal Master Plan, 251, 252

Coastal Protection and Restoration Authority (CPRA), 251

Coastal Wetlands Planning, Protection and Restoration Act (CWPPRA), 250–251

CoCoRaHS. *See* Community Collaborative Rain, Hail, and Snow Network

co-created projects, 25–26; definition of, 15, 25; educational value of, 224; establishment of team for, 37; full-participation approach in, 28; vs. other types of projects, 15, 26; publicity for, 45

collaboration, with organizations, in recruitment, 79–81

collaborative projects, 25–26; data-collection protocols in, *122,* 122–123; definition of, 25; full-participation approach in, 28; history of, 18; vs. other types of projects, 26

collared peccaries, 244

collection stage of data life cycle, 140, 142–147. *See also* data collection

collegial projects, 17, 25

Comissão Pró-Índio do Acre (CPI-AC), 238–246

Commerce, U.S. Department of, 66n4

communication of results. *See* reporting results

communication with citizen scientists: development of plans for, 45–46; in Nature's Notebook, 196–197; online, 102; in retention, 93, 96; before training, 114–115; at Urban Ecology Center, 216–217

communities of practice, 200

community building: in retention, 94–96; in training, 105

Community Collaborative Rain, Hail, and Snow Network (CoCoRaHS), 132, 142, 196

community leaders: as project team members, 85; in recruitment, 78–79

community science, 15, 219

complexity, of protocols, 41–42, 90

conferences, in volunteer retention, 94–95

consent, informed, 60–62

Constructing Scientific Communities, 22

contractual projects, 25

contributory projects, 25–27; vs. co-created projects, 15; definition of, 15, 25; educational value of, 224; geographic scale of, 18; publicity for, 45

Cooperative Extension programs, 2, 79

Cooperative Observer Program, of National Weather Service, 18, 20, 22

COPPA. *See* Children's Online Privacy Protection Act

Copyright Act, 69

copyright law, 68–71

Cornell Lab of Ornithology: Celebrate Urban Birds at, 212; development of tools at, 92–93; NestWatch at, 80. *See also* eBird

Cornell University, 211

cost-benefit analysis, 38

cost reduction, as improper motivation for citizen science, 1

CPI-AC. *See* Comissão Pró-Índio do Acre

CPP. *See* California Phenology Project
CPRA. *See* Coastal Protection and Restoration
    Authority
Creative Commons, 69–70
creative compilations, copyright of, 68–70
creative works, copyright of, 68–69
credibility, scientific, 90, 164, 208
CRIKT. *See* Citizens Researching Invertebrate
    Kritters Together
cross-validation, of data interpretation by citi-
    zen scientists, 27
crowdsourcing projects: ensuring quality data
    in, 134–135; motivations for participating in,
    38; planning and design of, 33; recruitment
    for, 83
crowned urchins, 262
CWPPRA. *See* Coastal Wetlands Planning,
    Protection and Restoration Act

Darwin, Charles, 7
data: copyright issues with, 68–71; definitions
    of, 138; degradation of, 141, *141*; evaluation of
    goals for, 175–176, 179, 181; impact of techno-
    logical advances on, 8, 22, 43; privacy policies
    for, 52, 66–67; privacy risk by type of, 67, *67*;
    qualitative, 48, 151; types of, 138
data analysis: in Amazonian biodiversity moni-
    toring, 242–243, 245; checklist and guide-
    lines for, 157, 158; in data life cycle, 141, 151;
    definition of, 151; meta-analysis, 150; plan-
    ning and completing, 42–43, 47–48; prepar-
    ing data for, 151; qualitative, 48, 151; strengths
    of citizen scientists in, 9; tools for, *144–146*;
    at Urban Ecology Center, 214–215
databases: planning for, 141–142, *143*; software
    for, *145*
data centers and repositories, 149, 150
data collection: best uses for citizen scientists
    in, 12–13; community building through
    group, 95; in contributory projects, 26; in
    data life cycle, 140, 142–147; geographic scale
    of, 8, 15, 18; historical, 17–23; importance of
    planning for, 119–120; land ownership and
    access rights in, 57–58; liability in, 63–64;
    methods of (*See* data-collection protocols); in
    models of participation, 26, 28; pros and cons
    of using citizen scientists in, 26; providing
    feedback on, 46–47; providing options for

different roles in, 88–90; quality of (*See* qual-
    ity); strengths of citizen scientists in, 8–9,
    12–13; time spent planning for, 119; tools for,
    142–147, *144–146*; training on (*See* training)
data-collection protocols, 119–135; in
    Amazonian biodiversity monitoring, 241–
    242, *243, 245*; in collaborative projects, *122*,
    122–123; complexity of, 41–42, 90; creating
    new plan for, 123–129; data entry methods in,
    131–133; designing, in development phase of
    projects, 41–42; ensuring data quality in,
    133–135, *134*; for experimental vs. observa-
    tional studies, 123–124; flexibility of, 88–90;
    levels of intensity of, 124–129, *126, 128*; in
    Nature's Notebook, 201–203, *202*; in Reef
    Check California, 258–264; sampling
    designs in, 129–131, *130*; testing and modify-
    ing, 44–45; training on (*See* training); at
    Urban Ecology Center, 212, *213*; using exist-
    ing, 121–123, 134, 150; volunteer feedback on,
    44–45, 96
data discovery, 140, 149–150
data documentation: in data life cycle, 147–148;
    legal requirements for, 52; metadata in, 148;
    in research life cycle, 65–67
data entry, tools and methods for, 131–133
data interpretation: in models of participation,
    27, 28–29; planning and completing, 47–48;
    pros and cons of using citizen scientists in,
    27; at Urban Ecology Center, 214–215
data life cycle, 139–151; analysis stage of, 141,
    151; collection stage of, 140, 142–147; defini-
    tion of, 139; description stage of, 140, 148;
    discovery stage of, 140, 149–150; integration
    stage of, 140–141, 150; in Nature's Notebook,
    207, *207*; overview of stages of, 139–141, *140*;
    planning stage of, 139, 141–142, *143*; preser-
    vation stage of, 140, 148–149; quality assur-
    ance stage of, 140, 147
data life cycle management (DLM): definition
    of, 138; scientific method compared to, 138,
    *140*
data management, 137–151; in Amazonian bio-
    diversity monitoring, 242, 245; checklist and
    guidelines for, 156, 157; in data visualization,
    151; definition of, 138; ensuring data quality
    through, 134; importance of, 137, 138–139,
    157–158; in Nature's Notebook, 207, *207*; as

nature centers, recruitment through, 81, 82
Nature Conservancy, 15, 89, 169, 238, 242
Nature's Notebook, 191–209; challenges facing,
    192–193, 193, 204–209; collaboration with
    education centers, 81; data quality in, 201–
    204; description of, 144, 191–192; group
    engagement in, 197–201; operational frame-
    work for, 191–192, 192; recruitment and
    retention in, 193–197
NCO. See National Coordinating Office
neighborhood recruitment, 78–79
NEON. See National Ecological Observation
    Network
NestWatch, 102
New Orleans, Hurricane Katrina in, 250
newsletters, 167, 197, 217
NIH. See National Institutes of Health
nonverbal cues, 108
North American Bird Phenology Program, 18
North American Breeding Bird Survey. See
    Breeding Bird Survey
North Branch Prairie Project, 89
North Branch Restoration Project, 89, 95
North Chagrin Nature Center, 232, 233
Norway, history of citizen science in, 17–18
NSF. See National Science Foundation
NWS-COOP. See National Weather Service
    Cooperative Observer Program

objectives. See goals
observational metadata, 65–66
observational studies: vs. experimental studies,
    123–124; strengths of citizen science in, 13
observing networks, 200–201
Ohio, North Chagrin Nature Center in, 232, 233
Ohmage, 144
OneDrive, 149
one-time events, training for, 101–102
online training, 102
open access journals, 164
open data and open access policies, 56, 68
OpenOffice, 145
opportunistic studies, strengths of citizen sci-
    ence in, 13
organizations: public participation as goal of,
    13; recruitment through, 79–81
orientation training, 101
ornithology. See bird research

outcomes evaluation, 42
*Oxford English Dictionary*, 7

Packard, Stephen, 89
Paleontological Resources Preservation Act, 58
paper data-entry forms, 132, 147
Paperwork Reduction Act (PRA), 54, 55
Partnership for Interdisciplinary Studies of
    Coastal Oceans (PISCO), 258–259
Patagonia, 252
Pattern Perception, 27
PEAR Institute, 180
peer review, 7
peer-reviewed publications: creating versions
    of, for volunteers, 165, 166, 167; reporting
    results in, 162, 163–164, 170
PenguinWatch, 27
perception, evaluation of goals for, 177, 180,
    181–182
performance-based assessment, 105
personal impacts. See individual impacts
phenology, definition of, 191. See also specific
    projects
Phenology Trail (Tucson), 81
photographs: aerial, 250, 252–255, 253, 254; his-
    torical, 19
pie charts or graphs, 154, 155
pirarucu, 247
PISCO. See Partnership for Interdisciplinary
    Studies of Coastal Oceans
Plankton Portal, 27
planning stage of data life cycle, 139, 141–142,
    143
planning stage of research life cycle, 54–62. See
    also project planning
plants: historical collections and records of,
    19–20; public interest in, 10
Plants of Concern (POC) program, 87–96, 88
*PLoS One* (journal), 150, 164
Pocock, Michael, 10
Point Blue Conservation Science, 169
policies, 51–72; data (See data policies); defini-
    tion of, 52; in dissemination phase of
    projects, 68–71; history of citizen science's
    role in, 20–21, 58; in launch phase of
    projects, 63–65; law in relation to, 52; on legal
    and ethical responsibilities, 52; in manage-
    ment phase of projects, 65–68; in measure-

ment phase of projects, 71; in planning and design of projects, 40, 52, 54–59; in preservation phase of projects, 65, 67–68; privacy, 52, 66–67

policymakers, reporting results to, 168–169, 170

PopClock campaign, 197

popular epidemiology, 78

populations, definition and use of term, 129n2

PostgreSQL, 145

PRA. *See* Paperwork Reduction Act

precision, of data, 133–134, *134*

preliminary results, 163, 165–166

presentations: reporting results in, 167, 168, *170*; in training, 105–108, 110

preservation. *See* data preservation

PRIMER, 265, *270*

Privacy Act of 1974, 54, *56*

privacy policies, 52, 66–67

privacy risk, by type of data, 67, *67*

private land, *56*, 57–58

process studies, strengths of citizen science in, 13

professional advisory committees, 90, 211, 213

professional groups, recruitment through, 80–81

professional scientists: in advisory groups, 90, 211, 213; as advocates for policy change, 168–169; in Driven to Discover, 225, 229–230; funding for, xiii; need for, xiii, 1; origins of, 17; quality of data collected by, 133; reporting results to, 162, 163–164, *170*; shortage of, xiii; in USA National Phenology Network, 208

program evaluation. *See* project evaluation

Project Budburst, 82, 142, *144*

project evaluation, 173–183; definition of, 173; development of goals for, 42, 173–177, *174*, *178*; of effects, 173, *174*, *175*, 181–182; formative, 42; front-end, 42; indicators of success in, 173, *174*, *175*, 177–180; planning for, 42; summative, 42

project goals and objectives: for data use, 175–176, 179, 181; defining, 40, 74, 75, 173–175; evaluation of, 42, 173–177, *174*, *178*; for learning, 176–177, *178*, 179–180, 181; managing multiple, 40; for perception and action, 177, 180, 181–182; training in support of, 99; in volunteer recruitment, 75–76, 85–86

project metadata, 65

project planning and design, 33–50; analysis and reporting phase in, 47–50; common principles of, 33; considerations before committing to, 36, *36*; development phase in, 41–45; first steps in, 36–40; framework for effective, 34–36, *35*, *53*; live phase in, 45–47; overview of steps in, 34, *35*, 53, *53*; volunteer motivations in, 77; volunteer recruitment in, 37, 77; for volunteer retention, 88–90. *See also* research life cycle

protocols. *See* data-collection protocols

*Pterygophora*, 266, *267*, 268, *269*

public, reporting results to, 167–168, *170*

public domain, data in, 70

public engagement: citizen science as part of, 2; in Driven to Discover, 226; in historical citizen science, 22; as organizational goal, 13; as project goal, 40

publicity, in live phase of projects, 45–46

Public Lab, 249–255

public land, *56*, 57, 57–58

purple urchins, 262, 266–268, *267*, *268*, 270

QA. *See* quality assurance

QC. *See* quality control

QGIS, 215, 253, 254

qualitative coding, 151

qualitative data, analysis of, 48, 151

quality, of data: accuracy and precision in, 133–134, *134*; ensuring, 133–135; evaluation of goals for, 175–176, 179, 181; in Nature's Notebook, 201–204; at Urban Ecology Center, 214

quality assurance (QA): in data life cycle, 140, 147; definition of, 147, 203; as goal of training, 100; in Nature's Notebook, *203*, 203–204; in Reef Check California, 259; reporting on use of, 164

quality control (QC): in data life cycle, 140, 147; definition of, 147, 203; as goal of training, 100; in Nature's Notebook, 203, 204; need for, 10, 12; in project planning, 43; reporting on use of, 164; types of, 10; at Urban Ecology Center, 214

R (software), *146*, 214–215, 216

rafting guides, 80–81

scientific expertise: of amateurs, 7–8, 17–18; need for, xiii; of project team members, 85

scientific literacy, of citizen scientists, xiii–xiv, 1–2, 22

scientific method, 119, 138, *140*, 227, *227*

scientists. *See* amateur scientists; citizen scientists; professional scientists

SciStarter, 83

scouting programs, 80

scuba divers. *See* Reef Check California

sea stars, 266, *267*, *268*, 270

sea star wasting syndrome, 266, 270

sea urchins, 262, 266–268, *267*, *268*, 270

self-reporting, 180

self-selection, 185

Sensr, *144*

Serengeti National Park, 73–75

Serra do Divisor region, 238

Shannon's diversity index, 119

Shirk, Jennifer L., 182

Significant Life Experience Research (SLER), 212

skills, of citizen scientists, changes in, 186

SLER. *See* Significant Life Experience Research

small-group discussions, in training, 110, 111, 112

SMART goals, 175

Smith, Bruce, 73

Snapshot Serengeti, 27, 73–75

Snapshot Wisconsin, 27

snowball recruitment, 79

social media: recruitment through, 82–83; reporting results on, 166, 168, *170*

social motivations, 76–77

software, for data collection, 142–147. *See also specific types*

Sokal, Robert R., *Biometry*, 131

Sonoma coast, Reef Check California on, 264–272, *265*, *267–270*

SOP. *See* standard operating procedure

SOS Amazônia, 238

space, training, 112–113

S-Plus, *146*

spreadsheets: planning for, 142, *143*; software for, *145*

SPSS, *146*

staff, in training, 113

stakeholders: of Amazonian monitoring program, 239–240; of USA National Phenology Network, 191, 204–205

stalked kelp, 266

standardization of data, 41, 121–122, 150

standard operating procedure (SOP), for data collection, 147. *See also* data-collection protocols

State Marine Conservation Areas, 258

State Marine Parks, 258

State Marine Reserves, 258

statistical analysis, *146*, 151

statistics, in sampling design, 129–131

*Statistics for People Who (Think They) Hate Statistics* (Salkind), 151

Stebbins Cold Canyon Natural Reserve, 110

Steffy, Gayle, 163

Stepenuck, Kris, 93

St. Hubert School, 232, 235–236

stratified random sampling, 129, *130*

students: benefits of citizen science to, 223–224; recruitment of, 74–75, 81–82; training of, 104; in Urban Ecology Center, 212

summative project evaluation, 42

sunflower sea stars, 266, *268*

SUNY Empire State College, Beetle Project at, 82

supporting materials: planning and development of, 44, 113, 114; types of, 44

Supreme Court, U.S., 69

SurveyMonkey, 102

surveys: design of, 41–42; online, 102

symbols, in data visualization, 152

Systat, *146*

systematic reviews, 150

systematic sampling, 129, *130*

tables: data management in, 141–142, *143*; data visualization in, 152, *153*

teachers: recruitment through, 82; training by, 104

team members: areas of expertise of, 85; recruitment of (*See* recruitment)

technological advances, impact on citizen science, 8, 22, 43

technological requirements, in project planning and design, 43–44

Ten Principles of Citizen Science, 33, 61

Founded in 1893,
UNIVERSITY OF CALIFORNIA PRESS
publishes bold, progressive books and journals
on topics in the arts, humanities, social sciences,
and natural sciences—with a focus on social
justice issues—that inspire thought and action
among readers worldwide.

The UC PRESS FOUNDATION
raises funds to uphold the press's vital role
as an independent, nonprofit publisher, and
receives philanthropic support from a wide
range of individuals and institutions—and from
committed readers like you. To learn more, visit
ucpress.edu/supportus.